Handbook on *Mycobacterium avium* subspecies *paratuberculosis*

P Das
Justin Davis K
Karthik K
Radhika Syam

PREFACE

Paratuberculosis or johns disease is an infectious disease of all ruminant animals caused by mycobacterium avium subsp paratuberculosis (MTP), considered by many to be the most serious infectious disease currently plaguing the worlds cattle, sheep and goat industry. At present, there exist no reliable means of diagnosis, treatment or prevention. These deficiencies have made control of this disease extremely difficult and as a result, the disease continues to spread amongst world's food animal industries at an alarming rate.

This handbook is intended as a general introduction to mtp for students and scientists who have been working in this field. This handbook has been fully revised and updated to reflect the many existing new developments and techniques in the field of handling and diagnosis of MTP. One main reason for separation of contents in to theoretical and practical is for better understanding and to create interest among the readers. The authors have presented a well illustrated text which the readers will find interesting as well as informative, with technical details.

In keeping with modern teaching, this handbook adopts a functional approach and emphasizes the classification, physiology, pathogenic interaction and practical significance of MTP.

I am grateful to the many people who allowed me to reproduce materials in this book. My task in writing this handbook was made easier by the availability of many recent reviews and publications. I take this opportunity to extend my sincere thanks to our director, indian veterinary research institute.

P DAS

INDEX

Chapter 1

Historical background of MAP infection

HISTORY

Mycobacterium avium subspecies *paratuberculosis* (MAP) is a bacteria first observed by Johne & Frothingham in 1895. By the early 1900's, pseudotuberculous enteritis was well recognized as a new disease and one that was widespread MAP causes paratuberculosis or Johne's disease (JD), an intestinal granulomatous infection. Scientists proposed a variety of alternative names for the disease, such as paratuberculosis and hypertrophic enteritis. In 1906, J. McFadyean coined the term "Johne's disease". O. Bang suggested that avian tuberculin could be used for diagnostic testing of animals. This was confirmed a year later in England by G.P. Male. Thus, even before the bacterium that caused Johne's disease was isolated, scientists believed it to be related in someway to the bacterium causing tuberculosis in birds with a difference was that the organism causing avian tuberculosis could be grown on laboratory culture but the organism causing Johne's disease could not [1].

A serendipitous observation by the British scientist F.W.Twort led to the isolation of the etiologic agent during attempts to culture the Johne's disease bacterium. Twort's failure to keep pace with cleaning laboratory glassware and his discriminating eye enabled him to note small bacterial colonies growing like satellites around larger colonies in old cultures he was preparing to discard. The larger colonies were contaminants of the common hay bacillus, *Mycobacterium phlei*. Suspecting that the *M. phlei* bacteria were providing some essential nutrient, Twort incorporated a heat-killed preparation of *M. phlei* into his culture medium. This new culture medium, he discovered, supported the growth of new acid-fast bacterium. He named it "*Mycobacterium enteriditis chronicae pseudotuberculosae bovis Johne*". In 1912, H. Holth also reported successful isolation of the cause of Johne's disease. Holth recognized the disease descriptions of Bang and simply called the organism he isolated the *paratuberculosis bacillus*. He failed to receive much recognition for

(1)

his discovery[2].

After the etiologic agent could be cultivated in the laboratory, antigens were obtained from it for diagnostic testing. They were used in skin testing (as was done for tuberculosis), and for assays to detect antibodies in serum samples using complement fixation and agglutination techniques. The man who first referred to the disease as Johne's disease coined the term "johnin" for the skin test antigen since the equivalent preparation for skin testing for tuberculosis was called tuberculin. The next several decades were devoted to evaluation of these and other diagnostic tests and to improvements in methods for laboratory cultivation of the organism.In 1923, the first edition of Bergey's Manual of Determinative Bacteriology was published and officially named the causative agent of Johne's disease *Mycobacterium paratuberculosis*. The new age of paratuberculosis research was ushered in with the discovery of a genetic element unique to *M. paratuberculosis*. This nucleotide sequence in the chromosomal DNA of the organism was simultaneously and independently discovered by Des Collins in New Zealand and the research team led by J.J. McFadden in England in 1989. The sequence was found to be an insertion element and was designated IS900. It was the first insertion element ever reported in mycobacteria. The importance of this discovery was its enabling the development of genetic tools for the detection of *M. paratuberculosis* without having to cultivate the bacterium on laboratory media, a process typically requiring 12 to 16 weeks.

First recognised in cattle, then in sheep and later in goats. Paratuberculosis is worldwide in distribution and is considered as a threat to the dairy sector, beef/meat industry and livestock trade[3] in many countries including India. Inspite of exhaustive research on diagnosis, control and prevention for more than 100 years, JD remains as a challenge to the veterinary profession[4.].

The disease is worldwide in distribution including India and spreading insidiously, inflicting heavy economic losses to cattle, sheep, and goat industry [5,6,7].

References:

1. Manktelow, B.; Hellstrom, J. (1979). The history of Johne's disease bovines, *New Zealand Veterinary Journal.* 27(3) p. 48

2. Chiodini, R. J. 1993. The history of paratuberculosis (Johne's disease). A review of the literature 1895 to 1992. Pp. 1– 658. International Association for Paratuberculosis, Inc., Providence, Rhode Island

3. OIE (2004). Manual of diagnostic tests and vaccines for terrestrial animals. Vol.1. Paris- France, http://www.oie.int.

4. Stephens, L. R. and Aukema, R. (1989). In Johne's disease: Current trends in research, diagnosis and management, edited by Miller, A. R. and Wood, P. R. CSIRO, Melbourne, p.9

5. Chiodini, R.J., Van Kruiningen and Merkal, R. S. (1984). Ruminant paratuberculosis (Johne's disease): The current status and future prospect. *Cornell Vet.,* 74: 218-262.

6. Paliwal, O.P. and Rajya, B.S. (1982). Evaluation of paratuberculosis in goats. Pathomorphological studies. *Indain J. Vet. Pathol.,* 6: 29 - 34.

7. Tripathi, B.N., Munjal, S.K. and Paliwal, O.P. (2002). An overview of paratuberculosis (Johnis disease) in animals. *Indian J. Vet. Pathol.,* 26(1&2) : 1-10.

Chapter 2
GENERAL CHARACTERISTICS

The mycobacteria are a large, varied group of organisms, which include some significant pathogens (such as *M.tuberculosis, M.bovis, and M.leprae*) and many free-living non-pathogens. An important sub-group that can survive outside an animal host is the *M.avium* complex; *Mycobacterium avium* subsp. *paratuberculosis* belongs to this group. They are slow growing and resistant to treatments with acid and alcoholic compounds. This is due to their strong cellular wall, with a high lipid composition. *Mycobacterium avium* form a large group of closely related mycobacteria which can be subclassified into *Mycobacterium avium* subspecies *avium* (*Maa*), *Mycobacterium avium* subspecies *silvaticum* (*Mas*), and *Mycobacterium avium* subspecies *paratuberculosis* (*Map*) [1]

Analysis of the rRNA genes (rDNA) of mycobacteria has resulted in the division of this genus into two separate clusters. These correspond to the traditional fast-growing mycobacteria, represented by nonpathogenic environmental isolates, and the slow-growing mycobacteria, containing most of the overt pathogens[2,3,4.]The rDNA gene copy number also reflects this division between fast- and slow-growing mycobacteria; fast-growing mycobacteria contain two sets of rRNA genes, whereas the slow growers, including M. paratuberculosis and M. avium, contain only one copy[5,6,7]

M. avium subsp. *paratuberculosis* possesses the properties of *M. avium* along with some additional features. The organism causes paratuberculosis, a chronic enteric disease in animals. The reference strain is strain ATCC 19698[8].

Microscopic and macroscopic characters

Mycobacterium avium subsp. *paratuberculosis* is a facultatively aerobic, acid-fast and weakly gram-positive bacillus of 0.5-1.5 µm length. The size, color, and texture of a colony of MAP are dependent in part on the type of bacteriologic medium on which it is cultivated. On Herrold's egg yolk agar medium, one of the most commonly used culture mediums in veterinary diagnostic laboratories, the colonies appear small, somewhat rough and off-white to yellow in color. Pigmented (yellow) strains have been reported in sheep. On Middlebrook agar medium without Tween 80 (a detergent that improves the growth rate) the colonies are very rough in appearance and resemble those of M. tuberculosis. With addition of Tween 80 the growth rate of MAP increases and it's colonial morphology becomes smooth and domed resembling that of M. avium subsp. avium (MAA).

The cell wall comprises an inner peptidoglycan, an intermediate asymmetric bilayer of long chain fatty acid (mycolic acid linked to arabinogalactone, and an outer layer of peptiodglycolipids (mycoside). Lipoarabinomannan (LAM), a membrane anchored glycolipid inserted into cell wall, is one of several immunomodulatory cell wall components[9]. It is a slow growing micro-organism and requires mycobactin as a source of iron for its growth [10,11]. Mycobactin is high molecular weight complexes that chelate iron for storage in the bacterial cell wall. MAP shares over 95% DNA homology with *Mycobacterium avium* subsp. *avium*. The complex mycobacterial cell wall is relatively impermeable and is rich in lipids which is not only responsible for its acid-fastness, but also for its resistance to the external environment and to its ability to survive within macrophages of infected animals [12].

The bacteria were found to be more resistant to physical and chemical factors as compared to other mycobacteria. It was more resistant to chlorine, low pH, high temperature and can even withstand pasteurization[13]. Preliminary studies demonstrated that MAP were equally susceptible to UV inactivation as other bacteria, however, UV light had minimal effect on MAP viability in soil spiked with the organisms [14]. MAP can survive for one year on pastures, though the factors like drying of soil, exposure to sunlight and change of pH shorten the survival time of MAP in soil [15].

Antigens of MAP

Map specific antigens for diagnostic testing or preventive therapy was searched for long time back and led to discovery of several proteins(Table). Many of these proteins have homology to other mycobacterial antigens, such as the GroES and GroEL proteins. The GroES antigens are highly antigenic, small (ca. 100-amino-acid), highly conserved heat shock proteins. Recently, the GroES antigens of *M. avium* and *M. paratuberculosis* have been cloned and sequenced[17].

Not surprisingly, the *M. avium* and *M. paratuberculosis* GroES coding sequences and deduced amino acid sequences are 100% identical to each other and are over 90% identical to the *M. tuberculosis* and *M. leprae* sequences, differing by only 3 amino acids.

Known immunogenic proteins of *M. paratuberculosis*[16]

M. paratuberculosis protein	Characteristic	Size (kDa)
GroES	Heat shock protein	10
AhpD	Alkyl hydroperoxide reductase D	19
32-kDa antigen	Fibronectin binding properties, secreted protein	32
34-kDa antigen	Cell wall antigen, B-cell epitope	34
34-kDa antigen	Serine protease	34
34.5-kDa antigen	Cytoplasmic protein, *M. paratuberculosis* species specific	34.5
35-kDa antigen	Immunodominant protein	35
42-kDa antigen	Cytoplasmic *M. paratuberculosis*-specific protein	42
44.3-kDa antigen	Soluble protein	44.3
AhpC	Alkyl hydroperoxide reductase C	45
65-kDa antigen	GroEL heat shock protein	65

Other immunoreactive proteins of *M. paratuberculosis* include a 32-kDa secreted protein with fibronectin binding properties implicated in protective immunity[18,19] and a 34-kDa cell wall antigenic protein homologous to a similar protein in *M. leprae*. This 34-kDa protein carries two species-specific B-cell epitopes that have been exploited in the histopathological diagnosis of Johne's disease [20]

Another strongly immunoreactive protein of 35 kDa has also been identified in *M. avium* complex isolates, including *M. paratuberculosis* [21]. A more thoroughly characterized protein of 65 kDa from *M. paratuberculosis* is a member of the GroEL family of heat shock proteins [19] The alkyl hydroperoxide reductases C and D (AhpC and AhpD) are also characterized immunogenic proteins of *M. paratuberculosis* [22] Unlike other mycobacteria, large amounts of these antigens are produced by *M. paratuberculosis* when the bacilli are grown without exposure to oxidative stress.

AhpC is the larger of the two proteins and appears to exist as a homodimer in its native form since it migrates at both 45 and 24 kDa under denaturing conditions. In contrast, AhpD is a smaller monomer, with a molecular mass of about 19 kDa. Antiserum from rabbits immunized against AhpC and AhpD reacted only with *M. paratuberculosis* proteins and not with proteins from other mycobacterial species, indicating that antibodies against these proteins are not cross-reactive. Furthermore, peripheral blood monocytes from goats experimentally infected with *M. paratuberculosis* were capable of inducing gamma interferon (IFN-) responses after stimulation with AhpC and AhpD, confirming their immunogenicity[22] In conclusion, these proteins are potentially useful for developing future vaccines and diagnostic assays or monitoring disease progression.

Genome

MAP was traditionally distinguished from the ubiquitous *M. avium* by its extremely slow growth in culture and its requirement for exogenous source of mycobactin for in vitro growth. A DNA insertion sequence, IS900, found in multiple copies in the genome which was considered unique for MAP [23,24,25,26]. Some recent reports indicate that IS900-like sequences may be presented in other closely related Mycobacterial species which may affect the specificity of PCR tests targeting IS900 [27,28]. The genome sequence was deposited in Gene Bank and released to the public in January, 2004. The genome sequence of MAP revealed 19 different IS elements in the K-10 bovine strain[28].

Olsen *et al.* (2004) discovered the ISMPaI element and showed three copies were present in the genome. Another study by Johnsen et al. (2005) observed two insertion sequences IS1311 and IS1245 but Bannantine (2005) reported that IS1245 could mistakenly be observed and absent in MAP genome. The MAP genome is a single circular chromosome of 4829, 781 base pairs with G+C content 69.37. Approximately 1.5% (72.2 kb) of the MAP genome consists of repetitive DNA like insertion sequences, multigene families and duplicated housekeeping genes[29]Various insertion sequences have been reported: IS900 has 14-18 copies, IS1311 in seven copies and ISMAV2 in three copies. The sequence analysis also revealed several insertion sequences, with no identifiable homologs in other mycobacteria e.g. ISMAP02 present in 6 copies and ISMAP04 present in four copies.

These newly discovered IS elements may be of particular interest for their use as specific diagnostic targets due to their absence in other mycobacteria[29]The other identified genes related to pathogenesis were SOdA, invA, inVB, ahpC, fapP, groES, hsp65, hsp70 etc.

Disease

MAP is responsible for johns disease in ruminants, clinical disease is characterised by a progressive, afebrile weight loss that leads to emaciation, brisket or submandibular oedema and poor coat quality (roughness, loss of pigmentation, alopecia, 'wool-slip'), despite maintaining a good appetite.

A major feature of the illness in cattle is chronic, intractable diarrhoea, although in some animals this may be intermittent. In contrast to cattle, diarrhoea is not a feature in small ruminants. This is probably due to their greater ability to reabsorb water in the large intestine, though in advanced cases the feces may become soft and unformed.

MAP is also attributed to the cause Crohn's disease in human being which are characterised by Chronic diarrhoea associated with abdominal pain, fever, anorexia, weight loss, and a right lower quadrant mass or fullness are the most common presenting features[30] with increasing concern about the transmission of infectious diseases from animal to man, attention has refocused on *Mycobacterium paratuberculosis* as a candidate organism in the aetiology of Crohn's disease.

Reference

1. Thorel M-F; Krichevsky M; Levy-Frebault VV. (1990) Numerical Taxonomy of Mycobactin- Dependent Mycobacteria, emended description of Mycobacterium avium, and description of Mycobacterium avium subsp.avium subsp.nov., Mycobacterium avium subsp.paratuberculosis subsp.nov., and Mycobacterium avium subsp.silvaticum subsp.nov. Int. J. System. Bacteriol. 1990; 40:254-260

2. Rogall, T., J. Wolters, T. Florh, and E. C. Böttger. 1990. Towards a phylogeny and definition of species at the molecular level within the genus *Mycobacterium*. Int. J. Syst. Bacteriol. 40:323-330.

3. Stahl, D. A., and J. W. Urbance. 1990. The division between fast- and slow-growing species corresponds to natural relationships among the mycobacteria. J. Bacteriol. 172:116-124

4. Wayne, L. G., and G. P. Kubica. 1986. The mycobacteria, p. 1435-1457. *In* P. H. A. Sneath, N. S. Mair, M. E. Sharpe, and J. G. Holt (ed.), Bergey's manual of systematic bacteriology, vol. 2. The Williams & Wilkins Co., Baltimore, Md.

5. Bercovier, H., O. Kafri, and S. Sela. 1986. Mycobacteria possess a surprisingly small number of ribosomal RNA genes in relation to the size of their genome. Biochem. Biophys. Res. Commun. 136:1136-41

6. Chiodini, R. J. 1989. The genetic relationship between *Mycobacterium paratuberculosis* and the *M. avium* complex. Acta Leprol. 7:249-251.

7. Chiodini, R. J. 1990. Characterization of *Mycobacterium paratuberculosis* and organisms of the *Mycobacterium avium* complex by restriction polymorphism of the rRNA gene region. J. Clin. Microbiol. 28:489-494

8. Brennan, P.J. and Nikaido (1995). The envelop of mycobacteria. *Annual Rev. Biochemist.*, 64: 29-63

9. Chiodini, R.J., Van Kruiningen and Merkal, R. S. (1984). Ruminant paratuberculosis (Johne's disease): The current status and future prospect. *Cornell Vet.*, 74: 218-262.

10. Clarke, C.J. (1997). The pathology and pathogenesis of paratberculosis in ruminants and other species. *J. Comp. Path.*, 116: 217-261.

11. Hines, M. E., Kreeger, J.M., Herron, A.J. (1995). Mycobacterial infections in animals: pathology and pathogenesis. *Lab. Anim. Sci.*, 45(4): 334-351.

12. Manning, E.J.B. and Collins, M.T. (2001). *Mycobacterium avium* subsp *paratuberculosis* : pathogen, pathogenesis and diagnosis. *Rev. Sci. Tech. Off. Epiz.*, 20(1): 133-150.

13. Millar, D. S., S. J. Withey, M. L. V. Tizard, J. G. Ford, and J. Hermon-Taylor. 1995. Solid-phase hybridization capture of low-abundance target

DNA sequences: application to the polymerase chain reaction detection of *Mycobacterium paratuberculosis* and *Mycobacterium avium subsp. silvaticum*. *Anal. Biochem.* 226:325–330.

14. Cousins, D. V., R. Whittington, I. Marsh, A. Masters, R. J. Evans, and P. Kluver. 1999. Mycobacteria distinct from Mycobacterium avium subsp. paratuberculosis isolated from the faeces of ruminants possess IS900-like sequences detectable by IS900 polymerase chain reaction: implications for diagnosis. *Mol. Cell. Probes* 13:431–442

15. Beth Harris and Raúl G. Barletta *Mycobacterium avium* subsp. *paratuberculosis* in *Veterinary Medicine Clinical Microbiology Reviews*, July 2001, p. 489-512, Vol. 14, No. 3

16. Green, E.P., Tizard, M.L., Moss, H.J., Thompson, J., Winterbourne, D.J., McFadden, J.J. and Hermon, T.J. (1989). Sequence and characterisation of IS900, an insertion element identified in a human Crohn's disease isolate of *mycobacterium paratuberculosis*. *Nucl. Acid. Res.*, 17: 9063-9073.

17. Cobb, A. J., and R. Frothingham. 1999. The GroES antigens of *Mycobacterium avium* and *Mycobacterium paratuberculosis*. Vet. Microbiol. 67:31-35

18. Andersen, P., D. Askgaard, L. Ljunqvist, J. Bennedsen, and I. Heron. 1991. Proteins released from *Mycobacterium tuberculosis* during growth. Infect. Immun. 59:1905-1910

19. El-Zaatari, F. A. K., S. A. Naser, L. Engstrand, C. Y. Hachem, and D. Y. Graham. 1994. Identification and characterization of *Mycobacterium paratuberculosis* recombinant proteins expressed in *E. coli*. Curr. Microbiol. 29:177-184

20. Clemens, D. L., and M. A. Horwitz. 1995. Characterization of the *Mycobacterium tuberculosis* phagosome and evidence that phagosomal maturation is inhibited. J. Exp. Med. 181:257-270

21. El-Zaatari, F. A. K., S. A. Naser, and D. Y. Graham. 1997. Characterization of a specific *Mycobacterium paratuberculosis* recombinant clone expressing 35,000-molecular-weight antigen and reactivity with sera from animals with clinical and subclinical Johne's disease. J. Clin. Microbiol. 35:1794-1799

22. Olsen, I., L. J. Reitan, G. Holstad, and H. G. Wiker. 2000. Alkyl hydroperoxide reductases C and D are major antigens constitutively expressed by *Mycobacterium avium* subsp. *paratuberculosis*. Infect. Immun. 68:801-808

23. Collins, D.M., Gabric, D.M. and De Lisle, G.W. (1989). Identification of a repetitive DNA sequence specific to *Mycobacterium paratuberculosis*. FEMS Microbiology news letters. 60: 175-178.

24. Vary, P.H., Andersen, P.R., Green, E., Hermon-Taylor, J. and McFadden, J.J. (1990). Use of highly specific DNA probes and the polymerase chain reaction to detect *Mycobacterium paratuberculosis* in Johne's disease. *J. Clin. Microbiol.*, 28: 933-937.

25. Bull, T.J., Pavlik, Hermon-Tylor, J. and Tizzarol, M.L. (1990). Study of FS900 loci in Mycobacterium avium subsp. Paratuberculsosis by multiplex PCR% screening in *Proc. 6th ICP*, Melbourne,pp 265-274.

26. Cousins, D.V., Williams, S.N., Hope, A., and Eamns, G.J. (2000). DNA fingerprinting of Australian isolates of *Mycobacterium avium* subsp. *Paratuberculosis* using IS *900* RFLP. *Aust. Vet. J.,* 78(3): 184-190.

27. Motiwala, A. S., Amonsin, A., Strother, M., Manning, E. J. B., Kapur, V. and Sreevatson, S. (2004). Molecular epidemiology of *Mycobacterium*

avium subsp. *paratuberculosis* isolates recovered from wild animal species. *Journal of Clinical Microbiology.,*42: 1703–1712.

28. Li, L. Bannantine, J.P., Zhang, Q. et al., (2005). The complete genome sequence of Mycobacterium avium subsp. Paratuberculosis. Proc. Natl. Acad. Sci., USA.

29. Lingling, Li, Bannantine, J.P., Zhang, Q., Amosin, A., Barbara J, Marg, David, A.H., Baneji, N., Kanjial S., and Kapur, V (2005). The complete genome sequence of *mycobacterium avium* subsp. *paratuberculosis*. PNAS 122(35): 12344-49.

30. The Merck Manual, 14th edition 1999, Merck Research laboratories, New Jersey, USA.

Chapter 3
Epidemiology of MAP

Epidemiology of Johne's disease is not well understood. Recent studies focused on transmission of infection, risk factors for infection, screening techniques and simulation modeling were inadequate. Novel genomic and proteomic techniques following the availability of full genome sequences of MAP could facilitate the identification of suitable testing targets indicative of early infection, which may ultimately help in better understanding of epidemiology of the infection [1].

Susceptible species

Other than cattle, sheep and goats, Johne's disease has been described in many species including deer, moose, antelopes [2], foxes, stoats and wild rabbits [3,4]. Besides these, MAP infection was also reported in other free living species such as rats, weasel, crow, jacksaw, wood mouse and badgers. Infected wild ruminants, can act as reservoir of infection of paratuberculosis for domestic animals. The monogastric animals like horses, mules, hogs, chicken and monkey's could be infected with MAP experimentally, but they remain asymptomatic shedders. However, there were no reports of natural infection in these species. Sheep and deer can become infected at any age and present clinical signs of the disease[5].

MAP has also been implicated with Crohn's disease (CD) in humans. It's symptomatic similarity with ruminant paratuberculosis and isolation of MAP from gut, lymphoid tissues and blood of CD patients suggested its role in the pathogenesis of this illness [6].

Transmission of infection

Faecal excretion is the main source of environmental contamination with MAP, which can spread infection between animals.

Exposure to organisms originating in faeces can occur by ingestion of contaminated pasture, soil, water or from faeces on the teats [7]. Clinically affected sheep can excret 108 organisms per gram of faeces measured by endpoint titration in Bactec culture [8]. Thus, even a single case may be sufficient for environmental contamination. Since, it has been shown that MAP survive for considerable periods (up to ~ 1 yr) in the environment, the level of contamination builds up over time. Subclinically infected animals can also shed MAP in faeces and infection may persist in a flock with few or no observed clinical cases [8]. There information on the culture of MAP from the sheep milk was not traceable. However, several studies had shown that subclinically infected cows could excret MAP in their milk. Streeter in 1995 isolated MAP organisms from colostrum (22%) and milk (3%) of subclinically infected cows [9]. Grant et al. (2000) detected MAP DNA is 88% of milk samples of sheep with positive gamma interferon (IFN-γ) test. In sheep enterprises generally it is believed that neonates remain with their dams and get infection via contaminated teats by oral exposure, so infection via milk as such has less practical significance.

Congenital infection

MAP organisms were isolated from foetuses of infected cows [10] and recently from sheep [11,12] which indicated possibility of intrauterine infection.

MAP infection via these routes was more frequent in animals with advanced stage of the clinical disease [12,13,]. MAP has been isolated from the semen of clinically affected bulls [14] and rams [15] but its significance has been hardly evaluated. No study has been carried out to asses the likelihood of either dam or foetus being infected with contaminated semen. However, the chance of transmission of infection by this route was believed to be extremely low[16].

World Prevalence

Johne's disease is primarily a disease of ruminants. It has been recognized as a problem in cattle for more than a century. It is present in all continents and spreads insidiously. Studies have shown high prevalence of Johnes disease in ruminants, particularly in cattle in several countries **Table 1:**

Worldwide prevalence of Johne's disease

Country	Prevalence rate (%)	Reference
UK	15 %	Clarke, 1997
USA	18 %	Clarke, 1997
Germany	14 %	Ebert *et al.*, 2007
Italy	19.2 %	Arrigoni *et al*, 2007
Brazil	9.2 %	Hori *et al.*, 2007
Austria	19 %	Khol *et al.*, 2005

Prevalence in India

The first case of paratuberculosis in undivided India was reported from Lahore [17] followed by another case in 1917 from a Military Dairy farm by Sheather (1918). Since then a large number of cases have been reported and many cattle farms have suffered huge economic losses due to this disease. The incidence of Johne's disease in India appear to reflect the scenario around the world

Prevalence estimates of JD in small ruminants and diagnostic tests used in India (CIRG, Project Report 2010)

Species	Prevalence (%)	Test	Region /State
Sheep	10.5	ZN	Avikanagar,Rajasthan
Sheep	16.0	DTH	Punjab

Goats	11.87	DTH	Maharastra
Sheep	24.83	DTH	Gujrat
Sheep	53.0	FC	Bikaner,Rajasthan
Goats	10.7	FC	Ranchi,Bihar
Sheep	7.97	FC	Izatnagar,UP
Goats	7.9	FC	Izatnagar,UP
Goats	15.2	S-ELISA	Ludhiana,Punjab
Sheep	49.0	T-PCR	Ajmer, Rajasthan
Goats	38.0	T-PCR	Ajmer, Rajasthan& Mathura,UP
Goats	19.6-31.8	S-ELISA	Western Rajasthan
Goats	30.9-50.0	S-ELISA	South and west UP

Prevalence of JD in large ruminants in India (CIRG, Project Report 2010).

Species	Prevalence (%)	Test	Region /State
Cattle	1.9-3.9	DTH	Haryana
Buffalo	6.5	DTH	Maharastra
Cattle	3.01	DTH	Malwa, MP
Cattle	2.8-13.33	DTH	Gujrat
Cattle	34.1	DTH	Pondicherry
Cattle	28.3	FC	Mathura, UP
Buffalo	48.0	TC	Agra, UP
Buffalo	70.0	T-PCR	Izatnagar, UP
Buffalo	40.0	T-PCR	Agra,UP
Buffalo	25.5-40.3	S-ELISA	South and West UP
Cattle	30.0-42.6	S-ELISA	South and West UP

Buffalo	23.3	S-ELISA	Punjab

ZN (Acid fast staining), FC (faecal culture), MC (milk culture), TC (tissue culture), S-ELISA (serum ELISA), T-PCR (tissue PCR), DTH (Johnin test)

Molecular epidemiology

Molecular techniques such as restriction fragment length polymorphism analysis of genomic DNA or IS900 PCR products demonstrated that sheep were usually infected with 'S' strains and cattle with 'C' strains [18,19].

The 'S' strains were difficult to isolate on routine media used for bovine isolates [20].'S' strains from UK were strongly yellow pigmented and mainly host adopted for sheep. Natural infection with 'S' strains has been occasionally reported in goats [21], cattle [22] and deer [23]. Bison (*Bison bison*) infected with different MAP strain which are distinct from cattle and domestic livestock by using PCR and restriction endonuclease analysis (REA) of IS1311 gene[24]. IS 1311 is an insertion sequence from *M. avium* and MAP which was used to distinguish between cattle and sheep strains of MAP. The IS1311 gene was found to be polymorphic at position 223 in the MAP isolates from cattle which distinguished isolates from sheep strain by PCR-REA technique.

Reference

1. Sergeant, E.S.G. (2005). Epidemiology of Johne's disease: Recent developments and future trends. Proc.of 8th ICP, pp. 587-595. (Chiodini et al., 1984)

2. Beard, P. M., Rhind, S. M., Buxton, D., Daniels, M. J., Henderson, D., Pirie, A., Rudge, K., Greig, A, Hutchings, M. R., Stevenson, K. and Sharp, J. M. (2001). Paratuberculosis infection of non-ruminant wild life in Scotland. *J. Comp. Pathol.*, **12**(4): 290-299.

3. Greig, A., Stevenson, K., Henderson, D., Perez, V., Hugues, V., Pavlik, I., Hines, M.E., McKendrick I. and Sharp, J.M. (1997). Epidemiological study of paratuberculosis in wild rabbits in Scotland. *J. Clin. Microbiol.*, **37**, 1746–1751.

4. Chiodini, R.J., Van Kruiningen and Merkal, R. S. (1984). Ruminant paratuberculosis (Johne's disease): The current status and future prospect. *Cornell Vet.*, **74**: 218-262.

5. Collins, M.T. (1996). Diagnosis of Paratuberculosis. *Vet. Clin. N. Amer.: Food animal practice*, 12, 357-351.

6. Whittington, R.J. and Sergeant, E.S.G. (2001). Progress towards understanding the spread, detection and control of *Mycobacterium avium* subsp. *paratuberculosis* in animal subpopulations. *Australian Vet. J.*, **79**: 267-278.

7. Whittington, R. J., Hope, A. F., Marshall, D. J., Taragel, C. A. Marsh, I. (2000). Molecular epidemiology of *Mycobacterium avium* subsp *paratuberculosis IS900* and *IS1311* restriction fragment length polymorphism analysis of isolates from animals and a human in Australia. *J. Clin. Microbiol.*, **38**: 3240-3248.

8. Streeter, R. N., Hoffsis, G. F., Bech-Nielsen, S., Shulaw, W. P. and Rings, D. M. (1995). Isolation of *Mycobacterium paratuberculsis* from colostrum and milk of subclinically infected cows. *Am. J. Vet. Res.*, **56**: 1322-1324.

9. Sweeney et al., 1992 R.W. Sweeney, R.H. Whitlock, A.N. Hamir, A.E. Rosenberger and S.A. Herr, Isolation of *Mycobacterium paratuberculosis* after oral inoculation in uninfected animals, *Am. J. Vet. Res.* **53** (1992), pp. 1312–1314.

10. Eppleston, J.; Whittington, R.J. (2001) Isolation of *Mycobacterium avium* subsp *paratuberculosis* from the semen of rams with clinical Johne's disease. *Australian Veterinary Journal.*, **79 (11)**: 776-777.

11. Lambeth, C. Reddacliff, L.A., Windsor, P. et al.(2004). Intrauterine and transmammary transmission of *Mycobacterium avium* subsp. *paratuberculosis* in sheep. *Australian Vet. J.,* **82:** 504-508.

12. Sweeney, R.W. (1996). Transmission of paratuberculosis. *Vet. Clin. N. Am. (Food Anim. Pract),* **12:** 305-312.

13. Larsen,A. B., Merkal, R.S. and Cutlip, R.C. (1975). Age of cattle as related to resistance to infection with *Mycobacterium Paratuberculosis. Am. J. Vet. Res.,* **35(3):** 367-369.

14. Eppleston, J.; Whittington, R.J. (2001) Isolation of *Mycobacterium avium* subsp *paratuberculosis* from the semen of rams with clinical Johne's disease. *Australian Veterinary Journal.,* **79 (11):** 776-777.

15. Buergelt, C.D., Donovan, G.A. and Williums, J.E. (2004). Identification of *Mycobacterium avium* subsp. *paratuberculsosis* by PCR in blood and semen of a bull with clinical paratuberculosis. *Intern. J. Appl. Res. Vet. Med.,* **2:** 130-134.

16. Twort, F.W. and Ingram, G.L.Y. (1912). A Method for isolating and cultivating *Mycobacterium enteritidis chronicae pseudotuberculosae* bovis johne and some experiments on the preparation of a diagnostic vaccine for pseudotububerculosae enteritis of bovines. *Proc. Royal Soc. Lond.* **84:** 517-543. (Cited from Clarke, 1997).

17. Bauerfeind, R.,S., Benazzi, R., Weiss, T., Schliesser, H., Williams and G. Balger. (1996). Molecular characterization of *Mycobacterium paratuberculosis* isolates from sheep, goats and cattle by hybridization with a DNA probe to insertion element IS*900. J. Clin. Microbiol.,* **34:** 1617-1621.

18. Cousins, D.V., Williams, S.N., Hope, A., and Eamns, G.J. (2000). DNA fingerprinting of Australian isolates of *Mycobacterium avium* subsp. *Paratuberculosis* using IS *900* RFLP. *Aust. Vet. J.,* **78(3):** 184-190.

(25)

19. Whittington, R.J., Reddacliff, L., Marsh, I. and Sauders, V. (1999). Detection of *M. a. paratuberculosis* in formalin fixed, paraffin-embedded intestinal tissues by IS900 polymerase chain reaction. *Aust. Vet. J.,* **77**: 392-397.

20. Collins DM, Gabric DM, de Lisle GW. 1990. Identification of two groups of *Mycobacterium paratuberculosis* strains by restriction endonuclease analysis and DNA hybridization. *J Clin Microbiol* 28:1591-1596.

21. Whittington, R. J., Taragel, C. A., Ottaway, S., Marsh, I., Seaman, J. and Fridriksdottir, V. (2001). Molecular epidemiological confirmation and circumstances of occurrence of sheep (s) strains of *Mycobacterium avium* subsp. _paratuberculosis in cattle in Australia and sheep and cattle in Iceland. *Vet. Microbiol.,* **79**: 311-322

22. de Lisle, G.W., Cannon , M.C., Yates, G.F., Collins, D.M. (2005). Abattoir surveillance of paratuberculsosis in farmed deer in New Zealand. Strains from sheep and cattle can be distinguished by a PCR test based on a novel DNA sequence differences. *J. Clin. Microbiol.,* **40:** 4760-4762.

23. Whittington, R.J., Marsh, I.B. and Whitlock, R.H. (2001b). Typing of IS 1311 polymorphism confirms that bison (bison bison) with paratuberculosis in Montana are infected with a strain of *Mycobacterium avium* subsp. *paratuberculosis* distinct from that occurring in cattle and other domesticated livestock. *Mol. Cell. Probes.,* **15(3):** 139-145.

Chapter 4

Microbe Host Interaction

The disease is transmitted to the young animals by contaminated feed and water with *M.a. paratuberculosis*. Bacteria are taken up by M cells in the mucosa of the Peyer's patch of the small intestine[1]. M cells have microvilli or microfolds on their apical surfaces and they adhere to adjacent cells by tight junctions and desmosomes. M cells are important in antigen sampling and its transportation to the immunocompetent cells of the underlying lymphoid tissue [1, 2]. In case of antigens overload, the bacteria can invade the epithelial enterocytes [3]. The bacilli cross M cells by transcellular or paracellular transcytosis mechanisms and are expelled on the basolateral side of M cells pockets, from where bacteria are phagocytosed by subepithelial macrophages or dendritic cells. By *in-vivo* [4] and *in-vitro* [5, 6] experiments, it was found that MAP also entered the mucosa by enterocytes.Wu *et al.* [7] employed surgical approach to characterize the very early stages of calf infection with MAP. They found that strains of MAP were able to traverse the intestinal tissues within 1 hr of infection to colonize distant organs such as liver and lymph nodes. Both ileum and mesenteric lymph nodes were persistently infected for months following intestinal deposition of MAP despite the lack of mycobacterial faecal shedding.

During the first 9 months of infection, the humoral immune responses were not detectable. The levels of cytokines detected in both ileum and lymph nodes indicated a Th1-type associated cellular responses but not Th2-type associated humoral responses.Fibronectin (FN) attachment proteins (FAP's) are family of fibronectin binding proteins present in several species of mycobacteria [8, 9].

The tropism of *M. a. paratuberculosis* to M cells has been found to be governed primarily by a FN dependent mechanism that involved the binding of fibronectin by fibronectin attachment protein expressed on *M. a. paratuberculosis* and by integrins (á5β1) present on the luminal surface of M cells. Disruption of FN binding protein, either by M cells or by *M. a. paratuberculosis* abolished the M cell targeting and markedly diminished the invasive potential of the organism[9].

Bannantine *et al.* [10] found out a new gene encoding the *M. a. paratuberculosis* 35 kDa major membrane protein (MMP). The gene was expressed at a higher level in low oxygen and high osmolarity conditions that were similar to environment of the intestine. The 35 kDa MMP was a surface exposed protein that played a role in the invasion of the epithelial cells and was considered is a virulence factor of *M. a. paratuberculosis* that might be important in the initiation of infection *in vivo*.

Survival inside macrophages

Bactericidal mechanisms of macrophage include destruction of mycobacteria by enzymatic degradation (e.g. lysozyme) or acid production in mature phagosomes or by production of reactive oxygen intermediates (super oxide, H_2O_2 etc.) and reactive nitrogen intermediates (nitric oxide) [11]. Growths of organisms were inhibited by restriction of available metabolites e.g. iron within the phagosome of the macrophages [12, 13] and due to increased release of nitric oxide (NO) [14].

Mycobacteria showed resistance to the effect of reactive oxygen and nitrogen metabolites [15] and presented induction of toxic oxygen molecules in macrophages using their complement receptor CR3 during initial phagocytosis. Ligation of this receptor with mycobacteria failed to initiate respiratory burst [16].

Macrophages are important antigen processing cells (APC) presenting mycobacterial antigens to T-lymphocytes, which stimulate specific adaptive immunity. Thus, mycobacteria are more readily killed by macrophages activated by cytokines (IFN-γ). Another protective strategy was apoptosis of the macropahge induced by activated lymphocytes causing release of mycobacteria, which were subsequently phagocytosed by other activated macropahges [16]. Activated macrophages were found to be stronger than ordinary macrophages and have many functions like ability to facilitate maturation of phagosomes to phagolysosomes, reduction of iron levels in phagosomes and increased release of NO [14, 16].

Maturation of macrophages also altered their morphology and converted themselves into more sturdy epithelioid and giant cells, which caused reduced survival and multiplication of MAP in the paucibacillary form of paratuberculosis [12].

Olesan *et al.* [17] reported that MAP organisms expressed high levels of the detoxifying enzymes alkyl hydroperoxide reductases C and D, which provided protection against reactive metabolites. *In vitro* study conducted shown that IFN-γ stimulated bovine macrophages did not produce sufficient NO to limit MAP multiplication [14].

Some demonstrated that infection of bovine macrophages with MAP caused down regulation of expression of both MHC class-I and class-II on the surface of macrophages, thus suppress antigen presentation to T-lymphocytes[18].

Inducible nitric oxide synthase (iNOS) is considered important in the control of mycobacteria, and could be marker of classic macrophage activation. Tuberculoid granulomas were associated with enhanced iNOS production.

Hypoxia and bacterial components including lipopolysaccharides and mycobacterial lipoarabinomanans will lead to iNOS expression in macrophages. In addition, iNOS was strongly iduced by Th1-type cytokines including IFN-γ, IL-2 and TNF-ἀ; therefore iNOS expression was a marker of Th1 mediated macrophage activation [19].

Stimutis et al. [20] tested the hypothesis that γδ T cells are capable of activating MAP infected macrophages. Peripheral blood derived macrophages were infected in vitro with live MAP, and autologous lymph node derived γδ T cells or CD4+ T cells were co-cultured with infected macrophages for 48 hr, at which time bacterial survival as well as production of nitrites and IFN-γ was evaluated. Incubation of MAP infected macrophages with autologous γδ T cells did not result in reduced intracellular bacterial viability compared to infected macrophage cultures without added T cells. IFN- γ production by infected cultures containing added γδ T cells was not enhanced compared to that of infected macrophages alone. This study suggested failure of antigen-stimulated γδ T cells and CD4+ T cells from sensitized cattle to upregulate nitric oxide and mycobactericidal activity of autologous MAP infected macrophages.

Gollnick et al. [21] reported that survival of MAP in bovine monocyte-derived macrophages is not affected by host infection status but depends on the infecting bacterial genotype. They investigated the ability of different MAP strains to survive in bovine monocyte derived macrophages (MDMs) of cows naturally infected with MAP and control cows. Following differentiation, MDMs were challenged in vitro with four MAP strains of different host specificity (cattle and sheep). Two hours and 2, 4 and 7 days after infection, ingestion, and intracellular survival of MAP strains were determined by fluorescence microscopy.

There was no effect of the origin of MDMs (JD positive or control animals) on phagocytosis, survival of bacteria, or macrophage survival. In contrast, important strain differences were observed. These findings suggested that some MAP strains interfere more successfully than others with the ability of macrophages to kill intracellular pathogens which may make it important to include strain typing when designing control programs.

Immunopathology

Innate immune response plays some role in the early stages of infection, which includes phagocytosis and attempted destruction of organisms by macrophages. Not only macrophages but also dendritic cells, NK cells, γδT cells, neutrophils and marginal zone B cells were important in the innate immune response to MAP [16]. The CD1 system was also involved in the innate immune response via antigen independent recognition and lysis of cells expressing CD1 by γδT cells and NK-T cells providing rapid response in early phases of infection [22].

The effective resistance to progressive infection of paratuberculosis has been associated with cell-mediated rather than humoral immunity and local rather than systemic responses in the early stages of infection [23]. Protection against mycobacterial diseases was reported to be mainly due to CD4+ Th1 cells that produced IFN- γ to activate macrophages, which then killed mycobacteria during phagocytosis. Cytotoxic CD8+ cells were considered to serve subsidiary function by releasing bacteria from infected cells, which could be killed during phagocytosis by activated macrophages. CD4+ T cells (MHC-II restricted) recognise antigens only when presented along with MHC-II molecules on the surface of APC, and CD8+ cells recognise antigen associated with MHC-I.

(33)

MHC-II molecules present antigens from within phagosomes in APC, whereas MHC-I molecules present antigens from within the cytoplasm of the APC. Thus MHC-I presentation in mycobacterial infection requires leakage of mycobacterial products from their phagosomal habitat into the macrophage cytoplasm [24, 25].

Role of lymphocyte subpopulation

Extensive studies have been made by various workers on the relative contributions of T-lymphocyte subsets to host defense in cattle infected with MAP.The host immune response to infection with MAP is paradoxical, with strong cell-mediated immune response during the early, subclinical stages of infection and strong humoral responses during the late clinical stages of the disease.

Cell mediated immune responses modulated by various T cell subsets are essential to provide protective immunity and prevent progression of the disease. Secertion of cytokines by T cell populations serves to activate macrophages to kill ingested MAP as well as activate other T cell subsets to contain the infection [26].

Activated T lymphocytes produce IL-2 which results in clonal expansion of specific CD8+ cytolytic T cells and CD4+ T helper cell populations. During activation of T cells, macrophages present antigen associated with MHC class II molecules to CD4+ cells. Antigen associated with MHC class I molecules is presented to CD8+ cells. The Th1 lymphocyte population produces IL-2, TNF-β, and IFN-γ, cytokines which direct cell-mediated immune function [27, 28, 29]. In contrast, the Th2 subpopulation of lymphocytes is responsible for induction of humoral immune function via the cytokines IL-4, IL-5, IL-6 and IL-10 [30, 31].

(34)

Bassey and Collins [32] measured the biological activity of each subset, expressed as lymphoproliferation and IFN-γ production, in response to phytohemagglutinin (PHA) and *M. avium* antigen preparation (A-PPD). The results showed a correlation between proliferation and IFN-γ production in response to A-PPD but not to PHA. In response to PHA, CD4+ lymphocytes were the most prolific producers of IFN-γ. CD8+ lymphocytes produced IFN-γ to a lesser extent, whereas γ δ+ T lymphocytes produced little or no IFN-γ.

Differences observed between the amount of IFN-γ produced by CD4+ versus CD8+ cells and CD4+ versus γδ+ cells were significant, but those between peripheral blood mononuclear cells (PBMC) and CD4+ T cells were not. Similar responses to A-PPD were observed except that PBMC produced higher levels of IFN-γ than did CD4+ T cells. These data for cattle was similar to observations made for other animal species, where CD4+ cells were the major type of T lymphocytes producing IFN-γ.

Role of cytokines

Molecular interaction between different immunocompetent cells, lymphocytes subpopulation and macrophages infiltrating the gut as a result of infection was based on the release of several proinflammatory and other cytokines in all mycobacterial infections including paratuberculosis (Friedland, 1993).

The effects of infection with MAP on cytokine production may influence immune regulation at the site of colonization, resulting in chronic inflammatory state associated with the latter stages of this disease. Among cytokines studied, IL-1α, IL-1β, IL-6 and IFN-γ were expressed significantly more in the infected animals than in noninfected control animals. The expression of TNF-α, however, was not different between the two groups of cattle [34].

(35)

The expression of IL-1β, TNF-α, RANTES (regulated upon activation, normal T-cell expressed and secreted), monocyte chemoattractant protein 1 (MCP-1) and IL-8 was lower in blood from cattle with subclinical paratuberculosis infection than in the uninfected cattle. The reduced TNF-α, RANTES and MCP-1 responses may weaken protective immunity and have a negative effect on granuloma formation and function [35].

The expression of the genes encoding IFN- γ, TGF-β, IL-5, IL-8, IL1-α, IL1-β, IL-6 in the ileal tissues from MAP infected cattle were greater than the expression in comparable tissues from uninfected cattle, while expression of gene encoding IL-16 was lower in tissue from infected cattle. Mesenteric lymph node draining the sites of infection expressed higher levels of IL1α, IL-8, IL-2 and IL-10 than similar tissues from the control animals. In contrast, gene encoding TGF-β and IL-16 were expressed at lower levels in the lymph nodes from infected cattle. The pattern of enhanced cytokine expression in ileal tissues from infected cattle is a dichotomy of classical Th1 (IFN- γ, IL1α and IL-6) and Th2 (IL-5) cytokine expression. Enhanced expression of mRNA for IL-8 was consistent with the large number of macrophage found at sites of infection [36].

IL-10 and TGF-β had inhibitory roles on the destruction of intracellular MAP, potentially through their effects on IFN- γ production. Therefore, upregulation of IL-10 and TGF-β even in the presence of high IFN- γ production (e.g. cells from naturally infected cows) likely resulted in less effective killing of the MAP in these animals compared to healthy animals [37]. Among the cytokines examined by reverse transcription-polymerase chain reaction (RT-PCR), Th2-type cytokines IL-4 and IL-10, and Th1-type cytokine IL-2 were expressed more significantly in the lepromatous group than in the tuberculoid and noninfected groups. No statistical differences were observed in the expression of IFN-γ, IL-1 β, TNF-α and GM-CSF among lepromatous, tuberculoid, and noninfected groups [38].

The expression of TGF- β1 (a cytokine known to have immunosuppressor effects), observed in macrophages and giant cells forming the lesions, was closely related to the number of MAP. In focal and multifocal forms, usually positive to IFN- γ test, bacilli were difficult to detect and TGF-β1 expression was low or absent. Diffuse multibacillary lesions, negative to IFN-γ, show large numbers of MAP and the highest percentage of immunolabelled cells. Diffuse paucibacillary forms, positive to IFN- γ, have low numbers of AFB and scant or no cells positive to TGF- β1. The high expression of TGF- β1 would be related to the inability of macrophages to limit the multiplication of MAP [39].

Clinical findings

In cattle, weight loss despite adequate rations, accompanied by chronic diarrhoea were standard clinical signs. Hypoproteinemia and 'bottle jaw' or dependent mandibular oedema, were also reported in cases of advanced disease. Diarrhoea was infrequently observed with paratuberculosis in sheep, goats, bison and perhaps other non-domestic hoofstock species [40]. The symptoms could be exacerbated by stress factors such as parturition, low plane of nutrition, concurrent diseases etc. [41]. Reduction in milk yield has also been found to be associated with MAP infection in dairy cows [42].

An experimental paratuberculosis study in cattle revealed decrease in the total serum protein, albumin, triglycerides (TRIG) and cholesterol, while enzyme activities for creatine kinase (CK), fructose-1-6-diphosphate, aldose (ALD), lactate dehydrogenase, aspartate amino transferase were elevated.

These biochemical indicators could be helpful in advanced cases in cattle but no serum biochemical test was found to be reliable in the diagnosis of subclinical paratuberculosis [43].

Gross lesions

On necropsy, the diseased animals showed emaciation, serous atrophy of fat deposits (gelatinization of fat), intermandibular oedema, and serous effusion into body cavities. In cattle primary lesions were limited to the gastrointestinal tract and regional lymph nodes [3]. The main macroscopical findings were of chronic enteritis, chronic intestinal lymphangitis and mesenteric lymphadenopathy [23]. Lesions were more prominent in the distal small intestine, particularly in the distal ileum and associated lymph nodes in the natural infection [44]. Intestinal mucosal surface had thick, broad, closely packed transverse folds, which gave the mucosa a corrugated appearance [45]. The mucosa between folds may be reddened by congestion or ulcerated and in some cases the intestinal mucosa may appear granular or diffusely opaque [41]. Lymphatic lesions included lymphadenitis and lymphangitis. The mesenteric lymph nodes were enlarged, oedematous, pale and had little corticomedullary distinction. Subserosal and mesenteric lymphatics were prominent, dilated and were corded with granulomatous nodules [23]. Lesions in other organs found mostly in liver, kidneys and lungs consisted of focal granuloma. Vascular disorders, mostly arteriosclerosis and endocardial and aortic calcification have been reported [41].

Microscopic lesions

Based on severity, histological lesions of paratuberculosis in naturally infected cattle were classified as mild, moderate and marked.

(38)

Cattle with 'mild lesions' showed a single Langhan's giant cell in the lamina propria of villi or in the paracortical zone of mesenteric lymph nodes. Epithelioid cells were occasionally identified. The animals with 'moderate lesions' showed several small groups of macrophages or several individual giant cells or both in the lamina propria of intestinal villi, submucosa, subcapsular sinus or in the paracortical zone of regional mesenteric nodes. 'Marked lesions' consisted of many macrophages and giant cells spread throughout the mucosa, submucosa, muscularis and serosal layers. Villous lacteals were prominent and dilated, and some contained a few inflammatory cells or some were ruptured. Crypt glands were distended and filled with neutrophils and mucoid substances. Peyer's patches were surrounded by inflammatory cells, but not infiltrated by them. The submucosa was widened either by infiltrating inflammatory cells or transudate [46]. Lately the lesion are classified or divided into five categories [47].

- Focal lesions - consisted of small granulomas in the ileal and jejunal lymph nodes or the ileocaecal lymphoid tissues. These granulomas were formed by macrophages with abundant lightly foamy, pale cytoplasm and large nuclei with sparse chromatin. In some cases, a few lymphocytes could be seen amongst the macrophages. Frequently, multinucleated Langhan's giant cells were found either in the granuloma or in isolated locations. Neither necrosis nor fibrous tissue was seen at the periphery of the granulomas.

- Multifocal type -small granulomas or scattered giant cells appeared in some intestinal villi, as well as in the lymph nodes. In the small intestine, these granulomas were surrounded by mild infiltrates of lymphocytes and plasma cells, located in some of the villi.

- Diffuse multibacillay lesions- associated with severe granuloamatous enteritis affecting different intestinal locations and lymph nodes, were formed by macrophages containing large numbers of acid-fast bacilli. In the lamina propria the macrophage infiltrate had a mosaic like appearance. Glands were widely separated due to infiltrates, and villi were frequently fused.

- Diffuse lymphocytic lesions - lymphocytes were the main inflammatory cells, with some macrophages or giant cells containing few if any mycobacteria.

- Diffuse intermediate forms - the infiltrate was formed by abundant lymphocytes and macrophages, and mycobacteria were present to varying degrees related to the number of macrophages. Clinical signs and gross lesions were mainly associated with diffuse forms.

In water buffaloes, the visible gross changes of intestinal thickening, mucosal corrugations and enlargement of mesenteric lymph nodes exhibited histological alterations characteristic of mild to moderate granulomatous inflammation. The histological lesions obsereved in these animals were

classified into 3 grades on the basis of type of cellular infiltration, granuloma formation, and presence of acid-fast bacilli [45].

- Grade-1 lesions - were marked by the presence of scattered epithelioid macrophages amid large number of lymphocytes in the intestinal villi and in the paracortical regions of the associated mesenteric lymph nodes.

- Grade-2 - revealed microgranulomas, infiltration with a large number of epithelioid macrophages besides lymphocyetes in the intestinal villi, and granulomas in the mesenteric lymph nodes.

- Grade-3 lesions - were characterized by the presence of epithelioid granulomas and giant cells in the intestines and the mesenteric lymph nodes.

- The Ziehl-Neelsen's stained tissue sections revealed acid-fast bacilli in grade-3 and grade-2 animals and acid-fast granular debris in grade-1 animals.

Reference

1. Momotani, E., Whipple, E., Theirmann, A. and Cheville, N. (1988). Role of M-cells and macrophages in entrance of *Mycobacterium paratuberculosis* into domes of ileal Peyer's patches in calves. *Vet. Path.*, 25: 131-137.

2. Sigurdardottir, O.g., Pier, C.M., and Evensen, O. (2001). Uptake of *Mycobacterium paratuberculosis* through the distal small intestinal mucosa in goats: an ultrastructural study. *Vet. Path.*, 38: 184-189.

3. Van der Giesson (1993). A molecular approach to the diagnosis and control of bovine paratuberculosis. Ph.D Thesis, University of Utrecht, The Netherlands.

4. Duraisamy (2006).Studies on early pathogenesis and dignosis paratuberculosis(Johne's disease) in sheep and goat. M.V.Sc thesis submitted to IVRI, Izatnagar.

5. Schleig, P.M., Buergelt, C.D., Davis, J.K., William, E., Monif, G.R.G. and Davidson, M.K. (2005). Attachment of *M.a.* subsp. *paratuberculosis* to bovine intestinal organ cultures. Method development and strain differences. *Vet. Microbiol.* 108: 271-273.

6. Sigurdardottir, O.G., Bakke-Mckellep, A.M., Djonne, B., Evensen, O. (2005). *Mycobacterium avium subsp. paratuberculosis* enters the small intestinal mucosa of goat kids in areas with and without Peyer's patches as demonstated with everted sleeve method. *Comp. immunol. Microbiol Infect. Dis.* 28: 223-230.

7. Wu, C., Livesey, M., Schmoller, S.K., Manning, E.J.B., Steinberg, H., Davis, W.C. Hamilton, M.J., and Talaat, A. M. (2007) Invasion and persistence of *Mycobacterium avium* subsp. *paratuberculosis* during early stages of Johne's disease in calves. *Infect. Immun.*, 75(5): 110-2119

8. Schorey, J.S., Li, Q., McCourt, D.W., Bang-Mastek, M., Clarkee Curtiss, J.E., Ratliff, T.L. and Brown, E.J. (1995). A *Mycobacterium leprae* gene encoding a fibronectin binding protein is used for efficient invasion of epithelial cells and schwann cells. *Infect. Immun.*, 63: 2652-2657.

9. Secott, T.E., Lin, T.L. and Wu, C.L. (2001). Fibronectin attachment protein homologue mediates fibronectin bidning by *Mycobacterium avium* subsp. *Paratuberculosis Infect. Immun.*, 69: 2652-2657.

10. Bannantine, J.P., Huntley, J.F., Miltner, E., Stabel, J.R., Bermudez, L.E. (2003). The *Mycobacterium avium* subsp. *paratuberculosis* 35 kDa protein plays a role in invasion of bovine epithelial cells. *Microbiol.*, 149: 2061-2063.

11. Cheville, N.F., Hostetter, J., Thomsen, B.V., Simutis, F., Vanloubbeeck.,Y. and Steadham, E. (2001). Intracellular trafficking of *Mycobacterum avium* subsp. *paratuberculosis* in macropahges. *Dtsch. Tierarztl. Wochenscher.* 108: 236-242.

12. Lepper, A.W. and Wilks, C.R. (1988). Intracellular iron storage and the pathogenesis of paratuberculosis. Comparative studies with other mycobacterial, parasitic or infectious conditions of veterinary importance. *J. Comp. Pathol.*, 98(1): 31-53.

13. Kurade, N.P. (1999). Studies on immunopathogenesis and enzyme linked immunosorbent assay based dignosis in ovine paratuberculosis. Ph. D. Thesis submitted to Indian Veterinary Research Institute, Izatnagar (U.P).

14. Zhao, B., Collins, M.T., Czuprynski, C.J. (1997). Effects of gamma interferon and nitric oxide on the interaction of *Mycobacterium avium* subsp. *paratuberculosis* with bovine monocytes. *Infect. Immun.*, 56(5): 1761-1766.

15. Gomes, M.S., Florido, M., Pais, T. F. and Appleberg, R. (1999). Improved clearance of *Mycobacterium avium* upon disruption of the inducible nitric oxide synthase gene. *J. Immunol.*, 162(11): 6734-6739.

16. Cooper, A. M., Appelberg, R. and Orme, I.M. (1998). Immunopathogenesis of *Mycobacterium avium* infection. *Front Biosci.*, 3: 141-148.

17. Olsen, I., Reitan, L.J., Holstad, G. and Wiker, H.G. (2000). Alkyl hydroperoxide reductases C and D are major antigens constituitively expressed by *Mycobacterium avium* subsp. *paratuberculosis*. *Infect. Immun.*, 68(2): 801-808.

18. Weiss, D.J., Evanson, O.A., McClenahan, C.J., Abrahamsen, M.S., Walcheck, B.K. (2001). Regulation of expression of major histocompatibility antigens by bovine macrophages infected with *Mycobacterium avium* subsp. *paratuberculosis* or *Mycobacterium avium* subsp. *avium*. *Infect. Immun.*, 69(2): 1002-1008.

19. Hostetter, J., Huffman, E., Byl, K. and Steadham, E. (2005). Inducible nitric oxide synthase immunoreactivity in the granulomatous intestinal lesions of naturally occurring bovine Johne's disease. *Vet. Path.*, 42(3): 241-249.

20. Simutis, F.J., Jones, D.E. and Hostetter, J.M. (2007). Failure of antigen-stimulated γδ T cells and CD4+ T cells from sensitized cattle to upregulate nitric oxide and mycobactericidal activity of autologous *Mycobacterium avium* subspecies *paratuberculosis*-infected macrophages. *Vet. Immunol. Immunopathol.*,116 : 1-12.

(43)

21. Gollnick, N.S., Mitchell, R.M., Baumgart, M., Janagama, H.K., Sreevatsan, S., Schukken, Y.H. (2007). Survival of *Mycobacterium avium* subsp. *Paratuberculosis* in bovine monocyte-derived macrophages is not affected by host infection status but depends on the infecting bacterial genotype. *Vet. Immunol. Immunopathol.* 120: 93-105.

22. Ulrichs, T. and Porcelli, S.A. (2000). CD1 proteins: targets of T cell recognition in innate and adaptive immunity. *Rev. Immunogenes.*, 2(3): 416-432.

23. Clarke, C.J. (1997). The pathology and pathogenesis of paratuberculosis in ruminants and other species. *J. Comp. Path.*, 116: 217-261.

24. Roitt,I., Brostoff, J., Male,D. (1998). Immunology, fifth edition, London, Mosby International Ltd., p.17.

25. Silva, C.L., Bonato, V.I., Lima, V.M., Faccioli, L.H., Leao, S.C. (1999). Characterization of the memory/activated T cells that mediate the long lived host response against tuberculosis after bacillus Calmette-Guerin or DNA vaccination. *Immunology.*, 97(4): 573-581.

26. Stabel, J.R. (2000). Cytokine secretion by peripheral blood mononuclear cells from cows infected with mycobacterium paratuberculosis. *Am. J. Vet. Res.*, 61: 754-60.

27. Billiau, A., opdenakar, G., van Damme, J., De Ley, M., Volckaert, G., Van Beeumen, J. (1985). IL-1: amino acid sequencing reveals microheterogeneity and relationship with an interferon inducing monokine. *Immunol. Today.*, 6: 235-236.

28. Lowenthal, J.W., Cerottini, J.C., MacDonald, H.R. (1986). IL-1 dependent induction of both IL-2 secretion and IL-2 receptor expression by thymoma cells. *J. Immunol.*, 137: 1226-1231.

29. Salgame, P., Abrams, J.S., Clayberger, C. (1991). Differing lymphokine profile of functional subsets of human CD4 and Cd8 T cell clones. *Science.*, 254: 279-282.

30. Cherwinski, H.M., Schumacher, J.H., Brown, K.D., Mosmann, T.R. (1987). Two types of mouse helper T cell clone. III. Further differenced in lymphokine synthesis between Th1 and Th2 clones revealed by RNA hybridization, functionally monospecific bioassays, and monoclonal antibodies. *J. Exp. Med.* 166: 1229-1244.

31. Mosmann, T.R. and Coffman, R.L. (1989). Th1 and Th2 cells: different patterns of lymphokine secretion lead to different functional properties. *Ann. Rev. Immunol.*, 7: 145-173.

32. Bassey, E.O.E. and Collins, M.T. (1997). Study of T-Lymphocyte subsets of healthy and *Mycobacterium avium* subsp. *paratuberculosis* infected cattle. *Inf. Immunity.*, 65(11): 4869–4872.

33. Friedland J.S., Robin J. Shattock and George E. Griffin. (1993). Phagocytosis of *mycobacterium tuberculosis* or particulate stimuli by human monocytic cells induces equivalent monocyte chemotactic protein-1 gene expression. Cytokine, 5(2):150-156

34. Lee, H., Stable, J. R., Kehrli, Jr. M. E. (2001). Cytokine gene expression in ileal tissues of cattle infected with *M. a paratuberculosis*. *Vet. Immunol. Immunopathol.*, 82: 73-85.

35. Buza, J.J., Mori, Y., Bari, A. M., Hikono, H., Aodon-geril, Hirayama, S., Shu,Y. and Momotani, E. (2003) *Mycobacterium avium* subsp. *paratuberculosis* infection causes suppression of RANTES, monocyte chemoattractant protein 1, and tumor necrosis factor alpha expression in peripheral blood of experimentally infected cattle. *Inf. Immunity.*, 71(12): 7223–7227.

36. Coussens, P. M., Verman, N., Coussens, M.A., Efftmen, M.D. and McNutty, A.M. (2004). Cytokine gene expression in PBMC and tissue of cattle infected with *M.a. paratuberculosis* evidence from an inherent proinflammatory gene expression patttern. *Infect. Immunol.*, 72: 1409-1422.

37. Khalifeh, M.S. and stable, J.R. (2004) Effects of gamma interferon, IL-10 and TGF- β on the survival of Mycobacterium avium subsp. Paratuberculosis in monocyte derived macrophages from naturally infected cattle. *Infect Immun.*,72(4): 1974-1982.

38. Tanaka, S., Sato, M., Onitsuka, T., Kamata, H. And Yokomizo, Y. (2005) Inflammatory cytokine gene expresson in different types of granulomatous lesions during asymptomatic stages of bovine paratuberculosis. *Vet. Path.*, 42: 579-588.

39. Muñoz, M., Delgado, L., Verna, A., Benavides, J., García-Pariente, C., Fuertes, M., Ferreras, M.C., García-Marín , J.F., Pérez, V. (2008). Expression of transforming growth factor-beta 1 (TGF-beta1) in different types of granulomatous lesions in bovine and ovine paratuberculosis. *Comp Immunol Microbiol Infect Dis. Jan.* (E-pub)

40. Manning, E.J.B. and Collins, M.T. (2001). *Mycobacterium avium* subsp *paratuberculosis* : pathogen, pathogenesis and diagnosis. *Rev. Sci. Tech. Off. Epiz.*, 20(1): 133-150.

41. Chiodini, R.J., Van Kruiningen and Merkal, R. S. (1984). Ruminant paratuberculosis (Johne's disease): The current status and future prospect. *Cornell Vet.*, 74: 218-262.

42. Beaudeau, F., Belliard, M., Joly, A. and Ceggers, H. (2007). Reduction in milk yield associated with *Mycobacterium avium* subsp. *paratuberculosis* (Map) infection in dairy cows. *Vet. Res.*, 38(4): 625-634.

43. Szillagyi, M., Kormendy, B., Suri, A., Tuboly, S., Nagy, G. (1989). Experimental paratuberculosis (Johne's disease) - studies on biochemical parameters in cattle. *Arch. Exp. Veterinarmed.* 43(3): 463-470.

44. Buergelt, C.D., Layton, A.W., Ginn, P.E. and Taylor, M (2000). The pathology of spontaneous paratuberculosis in the North American Bison (Bison bison). *Vet. Path., 37: 428-438.*

45. Sivakumar P., Tripathi, B. N. Singh, N. and Sharma, A. K. (2006). Pathology of naturally occurring paratuberculosis in water buffaloes (*bubalus bubalis*). *Vet Pathol.*, 43:455-462.

46. Buergelt, C.D., Hall, C., McEntee, K. and Duncan, J.R. (1978). Pathological evaluation of paratuberculosis in naturally infected cattle. *Vet Path* 15:196–207.

47. Gonzalez, J., Geijo, M.V., Garcia, C., Verna, A., Corpa, J.M., Reyes, L.E., Ferreras, M.C., Juste, R.A., Garcia, Marin, J.F., Perez, V. (2005). Histopathological classification of lesions associated with natural paratuberculosis infection in cattle. *J. Comp. Path.,* 133: 184-196.

Chapter 5

Diagnosis of MAP infection in Animals

The diagnosis of paratuberculosis in early and subclinical stages is difficult, but it is possible in the clinical stage [1,2,3]. The diagnostic tests are based on either detection of the organism and its genome or the immune response elicited by the host.

Conventional methods

Faecal smear examination

Direct acid-fast staining of faecal samples may reveal *M. a. paratuberculosis* organism in clinical cases, but the sensitivity of the method is low. The presence of acid-fast clumps in the smear is considered positive [4]. This is a useful method for detecting asymptomatic shedders [5].

Direct impression smear of tissue

Examination of Ziehl Neelsen stained impression smears of tissues collected at necropsy or through biopsy of ileum and mesenteric lymph node for the demonstration of acid-fast bacilli has been commonly used for the diagnosis of paratuberculosis but the detection of AFB in paucibacillary cases is difficult by this method. [4,3]

Isolation of the organism

Cultivation of *M. a. paratuberculosis* is the definitive diagnostic test for paratuberculosis [6,4,3]. The procedure is time consuming and requires 8-16 wks of incubation and use of specialized culture media [7].

The modified Herrold's egg yolk medium (HEYM) supplemented with mycobactin J and hexadecyl pyridinium chloride as decontaminant of tissue/faecal samples has been more frequently used than other egg based media for the primary isolation of *M. a. paratuberculosis* [8,7]. Other media such as modified Dubo's, Middlebrook 7H9, 7H10 and 7H11 and Lowenstein Jensen's have also been used.

(50)

Humoral immunity based methods

Various serological methods have been developed for the detection of antibody response elicited by the infection with *M. a. paratuberculosis*. Diagnostic test measuring antibody response lack sensitivity in early infections and hence their use have been restricted to the diagnostic confirmation of suspected clinical cases and herds for the absence of infection [9, 3].

Agar gel immunodiffusion test

The AGID test is considered as a reliable test for confirmation of paratuberculosis in clinically suspected cattle [10], sheep [11] and goats [12, 13]. The test is found to be highly sensitive as well as specific in clinically affected goats [10, 11, 14]. The most widely used antigen in the AGID is the protoplasmic extract of *M. a. paratuberculosis*.

Enzyme linked immunosorbent assay (ELISA)

At present ELISA is considered to be the most sensitive and specific test for detection of the humoral immune response to *M. a. paratuberculosis*. Different ELISAs such as unabsorbed, absorbed, lipoarabinomannan (LAM) and affinity purified antigen (APA)-ELISA have been developed [3, 4].

APA-ELISA using immunoaffinity purified antigen for the detection antibodies in goat sera against *M. a. paratuberculosis* was also developed lately [9].

Cell mediated immunity based assay

Johnin test

The intradermal johnin test, a delayed type of hypersensitivity reaction, is the only diagnostic test available at the field level, and it has been widely used for screening of paratuberculosis in small and large ruminants.

(51)

It is mostly used as a single intradermal test. An increase in the skin thickness over 2 mm after 72 hr after inoculation is regarded as positive.

Lymphocyte stimulation test (LST)

LST is a sensitive *in vitro* method for the detection of T-cell mediated immune response to different mycobacterial antigens. Hence, proliferative responses of lymphocytes either in peripheral blood or in the lymphocyte culture following stimulation with mycobacterial antigens have been utilized for the diagnosis of paratuberculosis [3].

Interferon gamma (IFN-γ) assay

The IFN-γ assay has been shown in a number of studies to be superior to humoral antibody tests in the detection of sub-clinical infection in both sheep [15, 16] and cattle [17, 18]. Sensitivity of this test in subclinical paratuberculosis in cattle ranged from 50-93%, when compared with repeated faecal culture results. However, clinically affected cattle may give negative results [18].

The expression of IFN-γ and IL-2 released by lymphocytes derived from blood, MLN and intestine were related to the types of histological lesions. Sheep with the diffuse paucibacillary lesions were commonly positive in IFN-γ and DTH tests (77 and 100%, respectively), whereas sheep with diffuse multibacillary lesions were less likely to be positive. In addition most of the sheep with lesions confined to the intestinal Peyer's patch (subclinical cases) were positive by IFN- γ test [15].

Histology based test

Indirect immunoperoxidase test

An indirect immunoperoxidase test for the histological diagnosis of paratuberculosis was described as an alternative method to ZN technique.

The method was sensitive particularly in cases in which there was no paratuberculosis lesions and and ZN technique failed to detect AFB [19, 20, 21, 22, 23]. An indirect immunoperoxidase procedure for the diagnosis of paratuberculosis where in formalin-fixed ileocaecal tissue containing large number of *M.a.paratuberculosis* was used as antigen and goat antibovine IgG labelled with peroxidase was used as a conjugate. A positive reaction was characterised by the presence of brownish granules in the cytoplasm of epitheloid macrophages in the lamina propria and submucosa [19].

Molecular techniques

Diagnosis of paratuberculosis by bacterial culture requires longer period, and detection of AFB in stained smears is a less sensitive and less specific method.

To overcome these problems, considerable efforts have been made into the development of sensitive method for the detection of MAP genome.

A genetic probe referred to as a deoxyribo nucleic acid (DNA) probe, employs a polymerase chain reaction (PCR) technique to determine the pres ence of the Johne's organism. The specificity of the combined culture and DNA probe tests approaches 100%. Polymerase chain reaction testing potentially offers one of the most sensitive methods for detection of the *M. paratuberculosis* infection because the pres ence of only one organism should provide a positive signal. Many PCR like nested PCR, semi-nested PCR, Real time (Q) PCR were developed for diagnosis purpose.

Recent trends

To meet out demand of rapid diagnostic assay for the diagnosis of Johne's disease, initiatives were taken to leap into the emerging era of nanodiagnostics.

The use of nanotechnoloogies for diagnostic application show great promise to meet rigorous demands of the clinical laboratory for sensitivity and cost-effectiveness.

Nanoparticles possess size dependent properties, particularly with respect to optical and magnetic properties, that can be manipulated to achieve a detectable signal. New nanodiagnostic tools include quantum dots, gold nanoparticles and centilivers. Nanoscale material like gold nanoparticle show considerable different properties from bulk metal as electrons undergo quantum-refinement and a high proportion of the atoms in small metal nanoparticles will be present on the surface.

One such property of gold-nanoparticle is surface plasmon resonance i.e., incident light can couple to plasmon excitation of the metal which involves light induced motion of all the valence electrons.

Plasmon-Plasmon resonance, resulting from the interaction of locally adjacent gold nanoparticles labels that have bound to a target, produces change in optical property i.e., characteristic red colour of gold colloid changes to bluish-purple because of this effect .The gold nanoparticle can be decorated with specific detector molecule using various conjugation chemistries and be used as diagnostic reagent

Lateral Flow Immunoassay

The lateral flow immunoassay (LFIA) or immunochromatographic strip (ICS), or dipstick test is one of the most attractive and widely used popular immunoassay methods in food and clinical diagnostic work today. In this assay the capture antibody is immobilized on the nitrocellulose membrane in a predefined position.

The detection antibody, coupled with colloidal gold or latex particles, is placed in an area near the sample application port. The absorbing blot membrane located at the opposite end of the sample application serves as the wick and facilitates fluid movement on the membrane. When a sample of bacteria, toxin, or antigen suspended in liquid is applied to the sample port of the device, it binds to the detection antibody conjugated to a gold or latex particle.

The antigen-antibody complex migrates laterally on the membrane by capillary action to the opposite end of the strip and through a porous membrane that contains two capture zones, one specific for the bacterial pathogen and another specific for unbound antibodies coupled to the gold or latex (control line).

The presence of only one (control) line on the membrane indicates a negative sample, and the presence of two lines indicates a positive result, which is visualized within 5–10 min [24, 25]. One drawback to this method is that it is less sensitive than the ELISA based assays and requires a bacterial cell concentration of about 10^7–10^9 cells for a positive reaction. In recent years, however, attempts have been made to improve the sensitivity of the assay by introducing an automatic reader, which avoids the ambiguity in reading the positive reactive bands with human eyes; or by introducing chemiluminescent-based detection of the antigen-antibody complex on the membrane.

Latex Agglutination (LA) and Reverse Passive Latex Agglutination (RPLA) Tests

LA- and RPLA-based commercial detection kits are the most rapid methods used for bacterial or toxin detection. Generally these methods require large amounts of antigen to show a positive reaction.

In these methods, antigen-specific antibodies are immobilized on latex particles and mixed with a sample in wells of microtiter plates. If the specific antigen is present in the sample in LA, a coagulated precipitate is observed.

In RPLA, a diffuse pattern will appear in the bottom; in its absence, a ring or button will appear and the latex does not play a role here—thus it is called a "passive latex agglutination test."

Enzyme-Linked Fluorescent Assay

Fluorescence-based detection by ELISA, called enzyme linked fluorescent assay (ELFA), became popular because of its improved sensitivity and quick results [25, 26].

In this assay, the enzyme (i.e., alkaline phosphatase) conjugated to the detection antibody breaks down the substrate (4-methyl umbellliferyl phosphate, MUP) to produce a fluorescent end product (methyl umbelliferyl), which can be sensitively detected by a spectrofluorometer. In a different format, a fluorophore molecule, instead of an enzyme, is attached to the detection antibody for direct interrogation of antigens or pathogens. The commonly used fluorescent molecules are rhodamine B, fluorecein isocyanate, and fluorescein isothiocyanate (FITC).

Time-Resolved Fluorescence Immunoassay

Time-resolved fluorescence immunoassay (TRFIA) is commercially marketed as dissociation enhanced lanthamide fluorescent immunoassay (DELFIA) by Perkin-Elmer Life Sciences (Akron, OH). A lanthamide chelate (europium, samarium, terbium, or dysprosium) is used as a label in the detection antibody. The method works similarly to ELISA in a microtiter plate, in which an antibody first captures an antigen which is then detected by using the lanthamide-labeled antibody.

A low pH enhancement solution is added to dissociate the label from the antibody after the reaction, and these free molecules rapidly form a stable new fluorescent chelate which can be read by a fluorescent reader. Unlike other fluorescent labels, lanthamide has a long fluorescence decay time and an exceptionally large Stokes' shift, thus can be read after the background noise has reduced.

Chemiluminescent Immunoassay

Chemiluminescent immunoassay (CLIA) has been developed to detect pathogens from samples. Detection antibodies conjugated to the chemiluminescent dyes such as 3(2-spiroadamantane) 4 methoxy 4(3-phosphoryloxy) phenyl 1,2 dioxetane (AMPPD), APS-5 (Gehring et al. 2004), and luminol (3 aminophthalhydrazide) enhanced with 4-iodophenol are shown to elicit a sensitive signal upon binding to target bacteria [27].

Electrochemical-Immunoassay

Electrochemical-immunoassay was adopted for bacterial detection as an alternative to a label-free assay system, in which the antigen-antibody reaction is measured amperometrically using the redox electrodes. In practice, electrochemical immunosensors are an extension of conventional antibody-based enzyme immunoassays (ELISA), in which catalysis of substrates by an enzyme conjugated to an antibody causes pH change, produces ions, or allows oxygen consumption that generates electrical signals on a transducer [28]. Amperometric, potentiometric, and capacitive transducers have been used for such applications. In amperometric detection, for example, alkaline phosphatase conjugated to an antibody hydrolyzes p-nitrophenyl phosphate to phenol, which is detected by voltammetry.

In light-addressable potentiometric sensors (LAPS), urease-conjugated antibody hydrolyzes urea, resulting in the production of carbon dioxide and ammonia that changes the pH of the solution. A silicon chip coated with a pH-sensitive insulator and an electrochemical circuit measures the alternating photocurrent as a light emitting photodiode shines on the silicon chip.

Biosensors

Surface Plasmon Resonance

Surface plasmon resonance (SPR) measures the changes in the refractive index resulting from binding of the antigen molecule to an immobilized antibody on the surface of metal films (Au or Ag) [29, 30]. The binding kinetics could be measured in as soon as a few seconds to a maximum of 15 min [31, 32].

Label-free, quantitative detection is the major advantage of SPR. Several commercial SPR instruments are currently available: Biacore (Biacore International SA, Switzerland), SPR-670M (Nippon Laser and Electronic Lab, Nagoya, Japan), Spreeta™ (Texas Instruments, Dallas, TX) [33, 34]. The sensitivity of most of the SPR-based methods is equivalent to an ELISA (10^5–10^8 CFU/mL).

Fiber-Optic Biosensors

Fiber-optic biosensors utilize the total internal reflection (TIR) property of

light when it travels through the waveguide and generates a boundary of evanescent waves on the surface of the waveguide. Antibody- or immunoassay-based fiber-optic biosensors provide increased sensitivity, selectivity, and speed compared to the conventional immunoassay techniques. [30, 35]

(58)

Antibody-coupled fluorescence wave guide biosensors are commonly employed in bacterial detection. In principle, a specific antibody is first covalently linked to the optic fiber that captures the bacteria of interest and a fluorescently labeled (e.g., Cy-5 or Alexa-Fluor 647) detection antibody binds specifically to the bacteria. When a 635-nm laser light is launched at the proximal end of the waveguide, fluorescence molecules are excited and generate an evanescent wave. Part of the emitted light energy is transmitted through the fiber and detected by a photodetector at wavelengths of 670–710 nm. Portable sensors, e.g., Analyte 2000 and RAPTOR, manufactured by Research International (Monroe, WA), are widely used for such applications. [36, 37, 38]

Antibody-Based Microfluidic Sensors

Research on microfluidics has burgeoned in the past decade from a fascinating concept to applications in clinical, molecular, biochemical, and medical diagnostics. The common materials used in the manufacture of microfluidic systems are silicon, glass, and polymers. The polymer, poly (dimethylsiloxane) (PDMS) has been widely used for designing immunoassays on chips. [39]

A microfluidic immunosensor promises to improve analytical performance by reducing the assay time and reagent consumption, increasing sensitivity and reliability through automation, and integrating multiple processes in a single device. [39, 40]

Protein/Antibody Microarrays

Microarrays were originally developed as a tool for genotyping and gene expression analyses.

(59)

The success of DNA and mRNA microarrays led to more promising ventures into targeting proteins, either toxins or cells using an antibody array format.

Unlike the nucleic acid arrays, in which the lysis of cells to release the DNA or RNA is required, protein microarrays have the advantage of detecting cells or toxins in one step. Although the protein microarray is a commercial success in medicine, it still remains in its infancy in respect to bacterial detection.

Mass Spectrometric Immunodetection

Matrix-assisted laser desorption/ionization (MALDI) is the most common technique used for mass spectrometric analysis of proteins using laser pulses. MALDI coupled with time-offlight (TOF) measures the mass of intact peptides. In recent years, MALDI-TOF has been applied to protein biochips, to study the interaction of recombinant antibody-antigen, and for direct detection of bacterial cells [41]. MALDI-TOF has been also shown to be an effective and rapid method for identification of whole bacterial cells [42, 43] and toxins [44].

μSERS Biochip Technology

μSERS is a novel label-free detection technology which uses surface-enhanced Raman scattering (SERS) microscopy. The chip comprises pixels of capture antibodies on a SERS active metal surface, which selectively binds the target bacteria in a sample. Using the Raman microscope, the SERS fingerprints are collected from the pixels on the chip. At each pixel, the bacteria are identified in the spectral domain by matching the unique SERS fingerprint against the library of known fingerprints. [45]

REFERENCES

1. Collins, M. T. (1996). Diagnosis of paratuberculosis. *Vet. Clin. North Am. Food Anim. Pract.* **12**:357-371.

2. McDonald, W.L., Ridge, S.E., Hope, A.F. and Condron, R.J. (1999). Evaluation of diagnostic tests for Johne's disease in young cattle. *Aust. Vet. J.*, **77**: 113-119.

3. Tripathi, B.N., Munjal, S.K. and Paliwal, O.P. (2002). An overview of paratuberculosis (Johnis disease) in animals. *Indian J. Vet. Pathol.*, **26(1&2):** 1-10.

4. Manning, E.J.B. and Collins, M.T. (2001). *Mycobacterium avium* subsp *paratuberculosis* : pathogen, pathogenesis and diagnosis. *Rev. Sci. Tech. Off. Epiz.*, **20(1):** 133-150

5. Paliwal, O.P. and Rajya, B.S. (1982). Evaluation of paratuberculosis in goats. Pathomorphological studies. *Indain J. Vet. Pathol.*, **6:** 29 - 34.

6. Chiodini, R.J., Van Kruiningen and Merkal, R. S. (1984). Ruminant paratuberculosis (Johne's disease): The current status and future prospect. *Cornell Vet.*, **74**: 218-262.

7. Whipple, D.L., Callihan, D.R. and Jarnagin, J.L. (1991). Cultivation of *Mycobacterium paratuberculosis* from bovine faecal specimens and a suggested standard procedure. *J. Vet. Diagn. Invest.*, **3**: 368-373.

8. Whipple, D.L. and Merkel, R.S. (1983). Modification in the techniques for cultivation of *Mycobacterium paratuberculosis*. *Proc. Int. Coll. Res. Paratuberculosis, Ames*, IA, pp. 82-92.

9. Rajukumar, K., Tripathi, B.N., Kurade, N.P. and Parihar, N.S., (2001). An enzyme linked immunosorbent assay using immuno-affinity-purification antigen in the diagnosis of caprine paratuberculosis and its comparison with conventional ELISAs. *Vet. Res. Commun.*, **25**: 539-553.

10. Sherman, D.M., Markham, R.J.F. and Bates, F. (1984). Agar gel immunodiffusion test for diagnosis of clinical paratuberculosis in cattle. *J. Am. Vet. Med. Assoc.*, **185**: 179-182.

11. Shulaw, W.P., Bech-Nieslon, S., Rings, D.M., Getzy, D.M. and Woodruff, T.S. (1993). Serodiagnosis of paratuberculosis in sheep by use of agar gel immunodiffusion. *Am. J. Vet. Res.*, **54**: 13-19.

12. Sherman, D.M., Markham, R.J.F. and Bates, F. (1983). Diagnosis of clinical and subclinical Johne's disease in goats using agar gel immunodiffusion (AGID) test. *Proc. Int. Collq. Res. Paratuberculosis Cattle. J. Am. Vet. Med. Assoc.*, **185**: 179-182.

13. Rajukumar, K., Tripathi, B.N., Kurade, N.P. and Parihar, N.S., (2001). An enzyme linked immunosorbent assay using immuno-affinity-purification antigen in the diagnosis of caprine paratuberculosis and its comparison with conventional ELISAs. *Vet. Res. Commun.*, **25**: 539-553.

14. Tripathi, B.N., Sivakumar Periasamy, Paliwal, O.P. and Singh, N (2006). Comparison of *IS900* tissue PCR, bacterial culture, johnin and serological test for diagnosis of naturally occurring paratuberculosis in goats. *Vet. Microbiol.*, **116**: 129-137.

15. Perez, V., Tellechea, J., Corpa, J.M.J., Gutierrez, M., Garcia, Marin, J.F. (1999). Relation between pathologic findings and cellular immune responses in sheep with naturally acquired paratuberculosis. *Am. J. Vet. Res.*, **58**:799-803.

16. Gwozdz, J. M., Thompson, K. G., Murray, A., Reichel, M, P., Manktelow, B. W. and West, D. M. (2000). Comparison of three serological tests and an interferon-□ assay for the diagnosis of paratuberculosis in experimentally infected sheep. *Aust. Vet. J.*, **78(11)**: 779-783

17. Billman-Jacob H., Carrigan, M., Cockram F. *et al.*, (1992). A comparison of interferon gamma assay with the absorbed ELISA for the diagnosis of Johne's disease in cattle. *Aust Vet J.*, **69(2)** : 25-28.

18. Stable, J.R. and Whittlock, R.H. (2001). An evaluation of a modified interferon-gamma assay for the detection of paratuberculosis in dairy herds. *Vet. Immunol. Immunopathol.*, **79(1):** 69-81.

19. Nguyen, H.T. and Buergelt, C., D. (1983). Indirect immuno peroxidase test for the diagnosis of paratuberculosis. *Am. J. Vet. Res.*, **44:** 2173-2174.

20. Navarro, J.A., Ramis, G., SEva, J., Pallares, F.J., Sanchey, J.,(1998). Changes in lymphocyte subsets in the intestine and mesenteric lymph nodes in caprine paratuberculosis. *J. Comp. Pathol.* **118** : 109-121

21. Tripathi, B.N. and parihar, N.S. (1996). Mycobacterial antigen detection by immunohistochemistry in sheep paratuberculosis. *Nat. Symp. Impact of Environmental Pollution on Health and Production of Livestock, Poultry and Wildlife,* Pantnagar 4-6 pp. 142-143.

22. Kurade, N.P. (1999). Studies on immunopathogenesis and enzyme linked immunosorbent assay based dignosis in ovine paratuberculosis. Ph. D. Thesis submitted to Indian Veterinary Research Institute, Izatnagar (U.P).

23. Kumar, A.A., Tripathi, B.N. and Julhe, D.K. (2006). Immunohistochemical demonstration of mycobacterial antigen in sheep experimentally infected with mycobacterium avium subsp. paratuberculosis. *Indian J. Vet. Pathol.,* **30(2):** 1-4.

24. Chapman, P.A. and Ashton, R. (2003). An evaluation of rapid methods for detecting *Escherichia coli* O157 on beef carcasses. *Int. J. Food Microbiol.* **87**: 279-285

25. Ray, B. and Bhunia, A.K. (2008). Fundamental Food Microbiology, Chapter 41. CRC press, Boca Raton, Florida

26. Vernozy-Rozand C, Mazuy-Cruchaudet C, Bavai C and Richard Y (2004). Comparison of three immunological methods for detecting staphylococcal enterotoxins from food. Lett. Appl. Microbiol. 39: 490–494

27. Zamora BM and Hartung M (2002). Chemiluminescent immunoassay as a microtiter system for the detection of *Salmonella* antibodies in the meat juice of slaughter pigs. J. Vet. Med. B. 49: 338–345

28. Warsinke A, Benkert A and Scheller FW (2000). Electrochemical immunoassays. Fresenius J. Anal. Chem. 366:622–634

29. Hsieh HV, Stewart B, Hauer P, Haaland P and Campbell R (1998). Measurement of *Clostridium perfringens* beta-toxin production by surface plasmon resonance immunoassay. Vaccine 16: 997–1003

30. Geng T and Bhunia AK (2007). Optical biosensors in foodborne pathogen detection. In: Knopf GK and Bassi AS (eds) Smart Biosensor Technology. Taylor and Francis, Boca Raton, Florida, pp 503–519

31. Dmitriev DA, Massino YS, Segal OL, Smirnova MB, Pavlova EV, Gurevich KG, Gnedenko OV, Ivanov YD, Kolyaskina GI, Archakov AI, Osipov AP, Dmitriev AD and Egorov AM (2002). Analysis of the binding of bispecific monoclonal antibodies with immobilized antigens (human IgG and horseradish peroxidase) using a resonant mirror biosensor. J. Immunol. Methods 261: 103–118

32. Lathrop AA, Jaradat ZW, Haley T and Bhunia AK (2003). Characterization and application of a *Listeria monocytogenes* reactive monoclonal antibody C11E9 in a resonant mirror biosensor. J. Immun. Methods 281:119–128

33. Homola J, Yee SS and Gauglitz G (1999). Surface plasmon resonance sensors: Review. Sens. Actuat. B-Chem. 54:3–15

(64)

34. Rich RL and Myszka DG (2006). Survey of the year 2005 commercial optical biosensor literature. J. Mol. Recognit. 19:478–534

35. Bhunia AK, Banada PP, Banerjee P, Valadez A, Hirleman ED (2007). Light scattering, fiber optic and cell-based sensors for sensitive detection of foodborne pathogens. J. Rapid Methods Automat. Microbiol. 15:121–145

36. Anderson GP, King KD, Gaffney KL and Johnson LH (2000). Multi-analyte interrogation using the fiber optic biosensor. Biosens. Bioelectron. 14: 771–777

37. Geng T, Morgan MT and Bhunia AK (2004). Detection of low levels of *Listeria monocytogenes* cells by using a fiber-optic immunosensor. Appl. Environ. Microbiol. 70: 6138–6146

38. Nanduri V, Kim G, Morgan MT, Ess D, Hahm BK, Kothapalli A, Valadez A, Geng T and Bhunia AK (2006) . Antibody immobilization on waveguides using a flow-through system shows improved *Listeria monocytogenes* detection in an automated fiber optic biosensor: RAPTOR™. Sensors 6: 808–822

39. Bange A, Halsall HB and Heineman WR (2005). Microfluidic immunosensor systems. Biosens. Bioelectron. 20: 2488–2503

40. Lim CT and Y Zhang (2007). Bead-based microfluidic immunoassays: The next generation. Biosens. Bioelectron. 22:1197–1204

41. Pavlickova P, Schneider EM and Hug H (2004). Advances in recombinant antibody microarrays. Clinica. Chimica. Acta. 343:17–35

42. Madonna AJ, Basile F, Furlong E and Voorhees KJ (2001). Detection of bacteria from biological mixtures using immunomagnetic separation combined with matrix-assisted laser desorption/ionization time-of-flight mass spectrometry. Rapid Commun. Mass Spectrom. 15: 1068–1074

(65)

43. Madonna AJ, Van Cuyk S and Voorhees KJ (2003). Detection of *Escherichia coli* using immunomagnetic separation and bacteriophage amplification coupled with matrix assisted laser desorption/ionization time-of-flight mass spectrometry. Rapid Commun. Mass Spectrom.17: 257–263

44. Nedelkov D and Nelson RW (2003). Detection of staphylococcal enterotoxin B via biomolecular interaction analysis mass spectrometry. Appl. Environ. Microbiol. 69: 5212–5215

45. Grow AE, Wood LL, Claycomb JL and Thompson PA (2003). New biochip technology for label-free detection of pathogens and their toxins. J. Microbiol. Methods 53: 221–233

Chapter 6

Immuinoprophylaxis for MAP infection

Mycobacterium avium subsp. *paratuberculosis* (MAP) is the causative agent of paratuberculosis, or Johne's disease. Being progressive, chronic and incurable enteritis, it is causing heavy economic losses to affected farms and livestock industries. This disease is more endemic in humid areas and due to unavailability of proper standardized diagnostic test world prevalence is unknown. *MAP* is transmitted horizontally via the fecal–oral route [Harris and Barletta, 2001] and vertically through the utero–placental route [Buergelt and Williams, 2003]. *MAP* can survive in the environment for periods of more than 1 year so hygienic measures at farm level are more important in control of this disease. [Harris and Barletta, 2001].

Pathogenesis, immune response and disease

Following oral ingestion, MAP passes through the epithelial barrier of the intestine. MAP can pass through M cells to the underlying Peyer's patches [Sigurdardottir *et al.*, 2005]. MAP is phagocytosed by subepithelial macrophages and evades immune elimination and modulates immune response but at the same time, some bacilli can be ingested and processed by professional antigen-presenting cells dendritic cells and prime a MAP-specific T-cell response [Kuehnel *et al.*, 2001].

Upon disease progression, an early proinflammatory Th1-like immune response considered to be protective eventually gives way to a predominant antibody-based Th2-like immune response in diseased animals.
IFN-γ -producing CD4⁺ T cells are key players in an effective immune response against intracellular mycobacteria such as MAP [Waters *et al.*, 2003]. CD8+ cytotoxic C lymphocytes may also exert an antimycobacterial effect through the secretion of cytokines, such as macrophage-activating IFN-γ

(69)

and TNF-α [Coussens, 2001]. TNF-α is a key cytokine in granuloma formation through release of chemotactic factors, leading to attraction of activated immune cells [Adams *et al.*, 1996]. Killed MAP vaccine can induce a strong antigen-specific IFN-γ response by CD4+ T cells from draining lymph nodes, but this vaccine was ineffective in decreasing MAP number in infected tissues [Simutis *et a.,l 2005*].

Infection of cattle with MAP can be subclinical and clinical, according to the severity of the symptoms and the immune response induced [Whitlock and Buergelt, 1996]. In sheep and goat, disease have similar symptoms as in cattle, with the exception that diarrhea is less frequent [Rideout *et al.*, 2003]. In certain animals, bacteria are cleared by the innate immune system and infection does not become established [Chiodini *et al.*, 1984]. Infected susceptible animals will remain asymptomatic carriers during the first 2–4 years after contact, while intermittently shedding very low numbers of bacteria in their feces. In this stage, cell-mediated immune responses are readily detected, but existing tests using purified protein derivative lacks specificity because of interference with environmental *Mycobacterium* spp.

In subclinical form of disease, the animal shows no clinical signs of Johne's disease, but infection by MAP can be detected by fecal culture owing to the intermittent MAP shedding in the feces.

A strong mycobacteria-specific cell-mediated immune response is developed during this stage but antibody responses are low [Harris NB and Barletta RG 2001]. In clinical form of disease, animals develop progressive disease characterized by intermittent or persistent diarrhea, gradual weight loss with normal food intake, reduced milk production and decreased fertility [Kennedy and Benedictus, 2001]. At this stage, most animals present antibodies against MAP and show persistent bacterial shedding in feces. Mycobacteria-specific Th1-type immune responses are weak at this clinical stage [Tiwari *et al.*, 2006].

(70)

Need for Vaccination for Johne's disease

In endemic areas, animals are exposed daily to MAP from the contaminated environment, but animals younger than 6 months, with a functionally immature immune system, are particularly susceptible to MAP [Larsen *et al., 1975*] and will become infected during the first months of life by ingestion of contaminated colostrum, milk, water and feed [Sweeney, 1996]. The immune system of the newborn has a strong tendency to adopt a Th2-type profile, whereas a Th1- type profile is thought to be essential for protection against intracellular pathogens, such as mycobacteria. Vaccination of dams may represent an alternative approach in the control of infection of newborns.

Vaccination against Johne's disease

Vaccination against MAP was first reported in 1926 by Vallée and Rinjard [Vallée and Rinjard 1926]. Their vaccine consisted of a live nonvirulent strain of MAP adjuvanted in a mix of olive oil, liquid paraffin and pumice powder.

During the 20th Century, a number of live-attenuated and killed whole-cell-based vaccines were developed both for bovine and ovine Johne's disease. Routinely, the vaccines, suspended in mineral oil, are inoculated subcutaneously in cattle within 30 days of birth in the brisket [Bakker *et al., 2000*]. In goats, sheep and deer, vaccines are generally injected in the neck behind the ear. Revaccination is not recommended [Gilmour, 1976].

A. Killed whole-cell-based vaccines

Strain 18 used in a commercial vaccine ("Mycopar") in the USA is actually composed of killed *M. avium* subsp. *avium* helps in increasing antibody response against MAP and prevented further infection but causes interference in ELISA results [Chiodini,1993; Spangler *et al., 1991*].

Strain ID-"Lelystad" vaccine, manufactured in Netherlands, is composed of heat-killed MAP bacteria suspended in a water-oil emulsion and Strain 316F –

"Gudair" vaccine developed in Spain are found to increase IFN-γ production [Corpa *et al.*, 2000; Reddacliff *et al.*, 2006]. MAP 5889 Bergey strain is an experimental vaccine developed in Hungary, composed of a heat-killed MAP but it gives erroneous results in delayed hypersensitivity test with purified protein derivative [Kormendy, 1994]. Killed commercial vaccine Strain 18 killed MAP field-isolate adjuvanted with human IL-12 induced strong local, systemic and enteric IFN-γ responses. [Park and Scott, 2001]. "Silirum" is a killed vaccine composed of MAP strain 316F combined with highly refined mineral oils to reduce the granuloma formation at the vaccination site. This vaccine developed cellular and humoral immune responses and will not interfere with Johnin test [Munoz *et al.*, 2005]. All these heat killed vaccines are given at less than 1 month of age.

Live-attenuated whole-cell-based vaccines

"Neoparase", an oil adjuvanted, freeze-dried live modified 316 F strain have been used as a therapeutic, post exposure vaccine with increased humoral and cell mediated responses but interferes with Johnin test [Gwozdz *et al.*, 2000]. "Paratuberkulose vaksine", is based on two British reference strains of MAP (2E and 316F) adjuvanted in a mix of olive oil, liquid paraffin and pumice powder [Saxegaard and Fodstad 1985]. These vaccines are administered after 1 year of age.

"AquaVax" is composed of an aqueous suspension of live MAP strain 316F induces low transient immune responses but confers very little protection after an experimental challenge [Begg and Griffin, 2005]. Cell wall competent or spheroplast MAP vaccines adjuvanted in either alum or saponin were not having significant effect on fecal shedding and lesions were present. Vaccines adjuvanted with saponin had less systemic side effects and was having weak interference with comparative skin tests than vaccines adjuvanted with alum. [Hines *et al.*, 2007]

All these whole-cell-based vaccines, particularly oil adjuvant ones interfere with the existing diagnostic tests for bovine tuberculosis as they induce a positive in vitro IFN-γ response and interfere with the existing sero-diagnostic tests for paratuberculosis and these are not really characterized with respect to their attenuation, making their use as marked vaccines impossible [Emery and Whittington 2004].

Subunit-based vaccines

The entire genome sequence of the K-10 strain of MAP has recently become available and provides a precious tool for the study of MAP antigens. The three members of the Ag85 complex are highly conserved proteins with mycolyl-transferase activity present in all mycobacterial species and abundantly secreted in mycobacterial culture filtrate. The Ag85A and Ag85B components of *M. tuberculosis* are among the most promising vaccine candidates for human tuberculosis [Andersen *et al.*, 2004].

Strong T-cell responses can be detected against Ag85 complex in low and medium shedder animals, but not in culture-negative cows [Shin *et al.*, 2005]. Heat-shock protein (Hsp) 65 (GroEL) and Hsp70 (DnaK) can also induce specific immune responses in MAP-infected and MAP-vaccinated cattle. As for PPD responses, the mycobacterial Hsp70-specific CMI responses decrease upon progression to the clinical stage of the disease [Koets *et al.*, 1999].

P22 is an MAP protein belonging to the LppX/LprAFG family of putative mycobacterial lipoproteins induces good IFN-γ and antibody [Rigden *et al.*, 2006]. Another lipoprotein, the 19-kDa (MAP0261c) protein stimulates strong humoral but weak IFN-γ production in infected cattle [Huntley *et al.*, 2005]. Superoxide dismutase (SOD) is a 23-kDa intracellular protein which strongly induces γ δ+ T cells in cattle [Shin *et al.*, 2005].

Alkyl hydroperoxide reductases C (AhpC) and D (AhpD) are constitutively expressed by MAP *in vitro* and homologous antigens elicited a strong IFN-γ response [Olsen *et al., 2000*].

Conclusion

An ideal vaccine against paratuberculosis or Johne's disease should provide sterile immunity, or abolish faecal shedding. The vaccines currently available reduce clinical symptoms but cannot avoid the contamination due to fecal shedding. A better understanding of the molecular and immunological processes involved in the progression to clinical paratuberculosis may help to develop more efficient vaccines. Finally, the development of subunit vaccines will require the further identification and discrimination of MAP antigens with either a strong immunodiagnostic potential. New experimental infection models to test vaccine efficacy are needed. Evaluation of the new vaccines should use experimental challenge conditions with dose and inoculation route similar to natural infection.

Due to the slow progression of the disease, a compromise must be found between the length of the follow-up to validate a potential protective efficacy and the cost-management involved in this study.

References

Adams JL, Collins MT and Czuprynski CJ. (1996). Polymerase chain reaction analysis of TNF-α and IL-6 mRNA levels in whole blood from cattle naturally or experimentally infected with Mycobacterium paratuberculosis. Can. J. Vet. Res. 60(4), 257–262

Andersen P, Gicquel B and Huygen K. (2004).Tuberculosis vaccine science. In: Tuberculosis Science (2nd Edition). Rom WN, Garay SM (Eds). Lippincott Williams & Wilkins, PA, USA 885–898

(74)

Bakker D, Willemsen PT and van Zijderveld FG. (2000).Paratuberculosis recognized as a problem at last: a review. Vet. Q. 22(4), 200–204

Begg DJ and Griffin JF. (2005). Vaccination of sheep against M. paratuberculosis: immune parameters and protective efficacy. Vaccine 23(42), 4999–5008

Buergelt C and Williams E. (2003). In utero infection of pregnant cattle by Mycobacterium avium subspecies paratuberculosis detected by nested polymerase chain reaction. Intern. J. Appl. Res. Vet. Med. 1, 4

Chiodini RJ, Van Kruiningen HJ and Merkal RS. (1984). Ruminant paratuberculosis (Johne's disease): the current status and future prospects. Cornell Vet. 74(3), 218–262

Chiodini RJ. (1993).Abolish Mycobacterium paratuberculosis strain 18. J. Clin. Microbiol. 31(7), 1956–8

Corpa JM, Perez V and Garcia Marin JF. (2000). Differences in the immune responses in lambs and kids vaccinated against paratuberculosis, according to the age of vaccination. Vet. Microbiol. 77(3–4), 475–485

Coussens PM. (2001). Mycobacterium paratuberculosis and the bovine immune system. Anim. Health Res. Rev. 2(2), 141–161

Emery DL and Whittington RJ. (2004).An evaluation of mycophage therapy, chemotherapy and vaccination for control of Mycobacterium avium subsp. paratuberculosis infection. Vet. Microbiol. 104(3–4), 143–155

Gilmour NJ. (1976). The pathogenesis, diagnosis and control of Johne's disease. Vet. Rec. 99(22), 433–434

Gwozdz JM, Thompson KG, Manktelow BW, Murray A and West DM. (2000).Vaccination against paratuberculosis of lambs already infected experimentally with Mycobacterium avium subspecies paratuberculosis. Aust. Vet. J. 78(8), 560–566

Harris NB and Barletta RG. (2001).Mycobacterium avium subsp. paratuberculosis in veterinary medicine. Clin. Microbiol. Rev. 14(3), 489–512 .

Hines ME, Stiver S and Giri D. (2007).Efficacy of spheroplastic and cell-wall competent vaccines for Mycobacterium avium subsp. paratuberculosis in experimentally-challenged baby goats. Vet. Microbiol. 120(3–4), 261–283

Huntley JF, Stabel JR and Bannantine JP. (2005). Immunoreactivity of the Mycobacterium avium subsp. paratuberculosis 19-kDa lipoprotein. BMC Microbiol. 5(1), 3

Kennedy DJ and Benedictus G. (2001).Control of Mycobacterium avium subsp. paratuberculosis infection in agricultural species. Rev. Sci. Tech. 20(1), 151–179

Koets AP, Rutten VP and Hoek A (1999). Heat-shock protein-specific T-cell responses in various stages of bovine paratuberculosis. Vet. Immunol. Immunopathol. 70(1–2), 105–115

Kormendy B. (1994).The effect of vaccination on the prevalence of paratuberculosis in large dairy herds. Vet. Microbiol. 41(1–2), 117–125

Kuehnel MP, Goethe R and Habermann A (2001). Characterization of the intracellular survival of Mycobacterium avium ssp. paratuberculosis: phagosomal pH and fusogenicity in J774 macrophages compared with other mycobacteria. Cell Microbiol. 3(8), 551–566

Larsen AB, Merkal RS and Cutlip RC. (1975).Age of cattle as related to resistance to infection with Mycobacterium paratuberculosis. Am. J. Vet. Res. 36(3), 255–257

Munoz M, Garcia Marin JF and Garcia-Pariente C (2005). Efficacy of a killed vaccine (Silirum) in calves challenged with MAP. In: Eighth International Colloquium on Paratuberculosis. Manning EJ, Nielsen SS (Eds). International Association for Paratuberculsois, Inc., Copenhagen, Denmark 208–217

Olsen I, Reitan LJ, Holstad G and Wiker HG. (2000).Alkyl hydroperoxide reductases C and D are major antigens constitutively expressed by Mycobacterium avium subsp. paratuberculosis. Infect. Immun. 68(2), 801–808

Park AY and Scott P. (2001).Il-12: keeping cell-mediated immunity alive. Scand. J. Immunol. 53(6), 529–532

Reddacliff L, Eppleston J, Windsor P, Whittington R and Jones S. (2006).Efficacy of a killed vaccine for the control of paratuberculosis in Australian sheep flocks. Vet. Microbiol. 115(1–3), 77–90

Rideout B, Brown S and Davis WC. (2003).Diagnosis and Control of Johne's Disease. The National Academies Press, Washington, DC, USA

Rigden RC, Jandhyala DM and Dupont C(2006). Humoral and cellular immune responses in sheep immunized with a 22 kilodalton exported protein of Mycobacterium avium subspecies paratuberculosis. J. Med. Microbiol. 55(Pt 12), 1735–1740

Saxegaard F and Fodstad FH. (1985).Control of paratuberculosis (Johne's disease) in goats by vaccination. Vet. Rec. 116(16), 439–441

Shin SJ, Chang CF and Chang CD. (2005).In vitro cellular immune responses to recombinant antigens of Mycobacterium avium subsp. paratuberculosis. Infect. Immun. 73(8), 5074–5085

Sigurethardottir OG, Valheim M and Press CM. (2004).Establishment of Mycobacterium avium subsp. paratuberculosis infection in the intestine of ruminants. Adv. Drug Deliv. Rev. 56(6), 819–834

Simutis FJ, Cheville NF andJones DE. (2005). Investigation of antigen-specific T-cell responses and subcutaneous granuloma development during experimental sensitization of calves with Mycobacterium avium subsp paratuberculosis. Am. J. Vet. Res. 66(3), 474–482

Spangler E, Heider LE, Bech-Nielsen S and Dorn CR. (1991).Serologic enzyme-linked immunosorbent assay responses of calves vaccinated with a killed Mycobacterium paratuberculosis vaccine. Am. J. Vet. Res. 52(8), 1197–1200

Sweeney RW. (1996).Transmission of paratuberculosis. Vet. Clin. North Am. Food Anim. Pract. 12(2), 305–512

Tiwari A, VanLeeuwen JA, McKenna SL, Keefe GP and Barkema HW. (2006).Johne's disease in Canada. Part I: clinical symptoms, pathophysiology, diagnosis, and prevalence in dairy herds. Can. Vet. J. 47(9), 874–882

Vallée H and Rinjard P. (1926).Etudes sur l'entérite paratuberculeuse des bovidés. Res. Gen. Med. Vet. 35(1)

Waters WR, Miller JM and Palmer MV(2003). eEarly induction of humoral and cellular immune responses during experimental Mycobacterium avium subsp. paratuberculosis infection of calves. Infect. Immun. 71(9), 5130–5138

Whitlock RH and Buergelt C. (1996).Preclinical and clinical manifestations of paratuberculosis (including pathology). Vet. Clin. North Am. Food Anim. Pract. 12(2), 345–356

Chapter 7

Impact on food industry and human health

Mycobacterium avium subsp. paratuberculosis (MAP) is causative agent of Johne's disease which is characterized by gradual wasting and chronic enteritis, resulting in decreased production, reduced fertility, and increased replacement rates (1). MAP can affect a variety of animals, most commonly domestic ruminants such as cattle, sheep and goats (2, 3). Other animals such as red deer (4), rabbits (5) and other nonruminant wildlife species (6) can also be infected by MAP. Johne's disease has been reported on every continent and its prevalence is increasing in many countries, particularly in dairy cattle. The digestive (intestinal) tract is considered the most common entrance route of MAP into the organism. Further development and multiplication of MAP in the intestinal mucosa depend on the natural immunity of animals and are determined by the ability of macrophages to reduce the intracellular multiplication of mycobacteria (7). It has been estimated that for every clinical case of Johne's disease existing in an infected herd, as many as four to eight other animals may have subclinical disease and be asymptomatic carriers of infection (8). A few infected animals develop clinical disease after a period of several years. In pre-clinical period, shedding of MAP is intermittent, while clinically affected cows may shed 10^6-10^8 CFU/g of fecal material. These pre-clinical animals can shed the organism periodically in both faeces and milk (9, 10). Such animals are real threat as it contaminates the farm environment, transmit disease to other animals or been sold to other farmer there for spreading the
disease.

MAP in Milk and Milk products

Cattle are the source of milk on the shelves of our supermarkets which are contaminated with MAP and Johne's disease in increasing numbers should be

(81)

a public health concern until the controversy over MAP as an etiologic agent of human disease is resolved. Milk could be contaminated with MAP by either direct shedding of the micro-organisms or faecal contamination during or after milking (11). MAP has been detected in raw and pasteurized milk from cows, sheep and goats and cheeses on retail sale in different countries. At the farm level, the sources of contamination of raw milk with MAP include direct shedding of the pathogen from clinically or subclinically infected cows, faecal contamination and mixing with contaminated milk in bulk tanks. There are many potential sources of microbiological contamination of dairy products during processing such as mixing of pasteurized and raw milk in the processing line, the addition of contaminated ingredients, or cross contamination along the processing chain (12). MAP was detected in milk for the first time in 1935, when it was isolated from three out of four milk samples from clinically ill cows (13, 14). This finding was confirmed by a number of further studies (15,14).

Furthermore, MAP has been isolated/ detected from udder tissue, supramammary lymph nodes and milk originating from cows with clinical signs of paratuberculosis. From asymptomaticcows, MAP has been isolated from supramammary lymph nodes, milk and from colostrum (9,10). It has been suggested that MAP-infected macrophages are present in lipid droplets on the cream layer of milk (16).

Publications dealing with the culture detection of MAP in milk and milk products have also been increasing in number over the last decade (17,18). The mechanism of the shedding of MAP organisms into milk has not yet been well investigated. Presumably, the shedding of MAP organisms into milk occurs by hematogenous or lymphatic spread. Colostrum is one of the earliest potential exposures of dairy calves to infectious agents like MAP (1).

MAP has been recovered from the colostrum of 22.2% of clinically normal but MAP infected cows (36% of heavy fecal shedders; 16% of light fecal shedders; (10).

Unpasteurized milk is available for consumers from farmers can be infected with MAP which gone unnoticed. Survivability of MAP after pasteurization is studied and reports a varied finding. Pasteurization performed at different temperatures and for different times was very effective in MAP devitalisation, resulting in a greater than 6 log 10 and 4 log 10 reductions in all 85% and 14% of pasteurized samples, respectively (19). Whilst, O'Doherty et al. (2002) found that all 396 samples of pasteurized milk examined were negative; the occasional occurrence of viable cells has been reported in other studies (21).

Millar et al. (1996) noted the presence of MAP DNA in pasteurized milk from a small-scale production unit, detected by the PCR method. In the study of Grant et al. (2005) MAP was isolated from milk samples (12 out of 27) after heat treatment at 72.5°C. It has been shown that viable MAP can be detected even after application of different levels of pressure in conjunction with pasteurization (24). MAP was detected in cheese available on the market (25,26). Both studies documented that higher detection rates of MAP can be obtained by PCR than by culture.

Nevertheless, PCR does not provide information about the viability of MAP cells. The screening of cheese found detectable quantities of MAP DNA.

Mycobacterial contamination in meat.

Johne's disease affects both dairy and beef cattle so, theoretically, meat could also be a potential food vehicle of transmission of MAP to humans. In the advanced stages of Johne's disease, MAP infection is likely to be widely disseminated throughout the animal including muscle, lymph nodes and blood.

It has therefore been suggested that meat from old dairy cows, which is used to make ground beef for human consumption, may represent a source of MAP infection for consumers in the USA (27). The major sources of microbial contamination in slaughterlines are: fleece, workers' hands, faecal pelletsand knife blades (28).

The only revealed study that specifically concerned with carcass contamination with MAP was by Meadus et al. (2008), in which MAP DNA was isolated from the surface of dressed carcasses which indicate that MAP like any other faecal borne pathogens can be removed or redistributed on the carcass surface by washing. MAP bacteria probably survive standard cooking temperatures. Mycobacterium paratuberculosis is the most heat-resistant mycobacterium present in retail beef (30). Even well-cooked meat may contain live paraTB. Studies show prolonged exposure to at least 74° C (165° F) may be necessary to eliminate the paratuberculosis bug. Mycobacterium paratuberculosis is also resistant to nitrites and the smoking process used in sausage production (31)

Conclusion

Reports suggest that MAP is contaminated in both milk and meat which are the main source of food from animals. But it is not sure whether meat and milk products contaminated with MAP expose the public to any risk of illness. As MAP could isolate from the patients suffering from Crohn's disease, some scientists argue that it can be one of the causative organism for the disease. MAP can be detected in the human body, and researchers and clinicians have published this fact in a number of case reports in the scientific literature. Moreover, persons with Crohn's disease are seven-fold more likely than the general population to have MAP associated with their disease (32).

However, there has been no definitive causal relationship established between MAP and a specific disease process in humans similar to that which has been established in animals. If MAP is transmitted from infected livestock to humans through the consumption of contaminated food, it is imperative that to prevent contamination of food from MAP.

References

1. Chiodini, R.J., H.J. Van Kruiningen, R.S. Merkal. 1984. Ruminant paratuberculosis (Johne's disease): the current status and future prospects. *Cornell Vet* 74:218-262.

2. Cocito, C., Gilot, P., Coene, M., de Kesel, M., Poupart, P. and Vannuffel, P. (1994) Paratuberculosis. Clin Microbial Rev 7, 328–345.

3. Collins M T (2003) Update on paratuberculosis: 1. Epidemiology of Johne's disease and the biology of *Mycobacterium paratuberculosis*. *Irish Veterinary Journal* 56 565–574.

4. Godfroid J, Boelaert F, Heier A, Clavareau C, Wellemans V, Desmecht M, Roels S and Walravens K (2000) First evidence of Johne's disease in farmed red deer (*Cervus elephas*) in Belgium. Veterinary Microbiology 77 283–290.

5. Greig A; Stevenson K; Henderson D; Perez V; Hughes V; Pavlik I; Hines ME; McKendrick I; Sharp JM. (1999) Epidemiological study of paratuberculosis in wild rabbits in Scotland. J. Clin.Microbiol.; 37: 1746-1751.

6. Beard PM; Henderson D; Daniels M; Pirie A; Buxton D; Greig A; Hutchings MR; McKendrick I;Rhind S; Stevenson K; Sharp M. (1999) Evidence for paratuberculosis in fox (vulpes vulpes) and stoat (mustela erminea). Vet. Rec., 145, 612-613

7. Bendixen P.H., Bloch B., Jorgensen J.B. (1981): Lack of intracellular degradation of Mycobacterium paratuberculosis by bovine macrophages infected in vitro and in vivo: light microscopic and electron microscopic observations. American Journal of Veterinary Research, 42, 109–113.

8. Council for Agricultural Science and Technology (2001) Johne's disease in cattle. Issue Paper No. 17 (May 2001). Ames, Iowa: Council for Agricultural Science and Technology.

9. Sweeney R W, Whitlock R H and Rosenberger A E (1992) Mycobacterium paratuberculosis cultured from milk and supramammary lymph nodes of infected asymptomatic cows. Journal of Clinical Microbiology 30 166–171.

10. Streeter R N, Hoffsis G F, Bech-Nielsen S, Shulaw W P and Rings M (1995) Isolation of Mycobacterium paratuberculosis from colostrum and milk of subclinically infected cows. American Journal of Veterinary Research 56 1322– 1324.

11. Grant I R, Ball H J and Rowe M T (2002) Incidence of Mycobacterium paratuberculosis in bulk raw and commercially pasteurised cows' milk from approved dairy processing establishments in the United Kingdom. Applied and Environmental Microbiology 68 2428–2435.

12. Pearce L E, Truong H T, Crawford R A, Yates G F, Cavaignac S and De Lisle G W (2001) Effect of turbulent-flow pasteurization on survival of Mycobacterium avium subsp. paratuberculosis added to raw milk. Applied and Environmental Microbiology 67 3964–3969

13. Alexejeff-Goloff N.A. (1935): Journal of Comparative Pathology, 48, 81.

14. Taylor T.K., Wilks C.R., McQueen D.S. (1981): Isolation of Mycobacterium paratuberculosis from the milk of a cow with Johne's disease. The Veterinary Record, 109, 532-533.

15. Smith H.W. (1960): The Examination of milk for the presence of Mycobacterium johnei. Journal of Pathology and Bacteriology, 80, 440–442.

16. Koenig G.J., Hoffsis G.F., Shulaw W.P., Bech-Nielsen S., Rings D.M., St Jean G. (1993): Isolation of Mycobacterium paratuberculosis from mononuclear cells in tissues, blood, and mammary glands of cows with advanced paratuberculosis. American Journal of Veterinary Research, 54, 1441–1445.

17. Rademaker J.L., Vissers M.M., Te Giffel M.C. (2007): Effective heat inactivation of Mycobacterium avium subsp. paratuberculosis in raw milk contaminated with naturally infected feces. Applied and Environmental Microbiology, 73, 4185– 4190.

18. Stephan R., Schumacher S., Tasara T., Grant I.R. (2007): Prevalence of Mycobacterium avium subspecies paratuberculosis in Swiss raw milk cheeses collected at the retail level. Journal of Dairy Science, 90, 3590–3595.

19. McDonald W.L., O'Riley K.J., Schroen C.J., Condron R.J. (2005): Heat inactivation of Mycobacterium avium subsp. paratuberculosis in milk. Applied and Environmental Microbiology, 71, 1785–1789.

20. O'Doherty A., O'Grady D., Smith T., Egan J. (2002): Mycobacterium avium subsp paratuberculosis in pasteurized and unpasteurized milk in the Republic of Ireland. Irish Journal of Agricultural and Food Research, 41, 117–121.

21. Grant I.R., Rowe M.T., Dundee L., Hitchings E. (2001): Mycobacterium avium ssp. paratuberculosis: its incidence, heat resistance and detection in milk and dairy products. International Journal of Dairy Technology, 54, 2–13.

22. Millar D., Ford J., Sanderson J., Withey S., Tizard M., Doran T., Hermon-Taylor J. (1996): IS900 PCR to detect Mycobacterium paratuberculosis in retail supplies of whole pasteurized cows' milk in England and Wales. Applied and Environmental Microbiology, 62, 3446– 3452.

23. Grant I.R., Williams A.G., Rowe M.T., Muir D.D. (2005): Efficacy of various pasteurization time-temperature conditions in combination with homogenization on inactivation of Mycobacterium avium subsp. paratuberculosis in milk. Applied and Environmental Microbiology, 71, 2853–2861

24. Donaghy J.A., Linton M., Patterson M.F., Rowe M.T. (2007): Effect of high pressure and pasteurization on Mycobacterium avium ssp. paratuberculosis in milk. Letters in Applied Microbiology, 45, 154-159.

25. Ikonomopoulos J., Pavlik I., Bartos M., Svastova P., Ayele W.Y., Roubal P., Lukas J., Cook N., Gazouli M. (2005): Detection of Mycobacterium avium subsp. paratuberculosis in retail cheeses from Greece and the Czech Republic. Applied and Environmental Microbiology, 71, 8934–8936

26. Clark D.L. Jr., Anderson J.L., Koziczkowski J.J., Ellingson J.L. (2006): Detection of Mycobacterium avium subspecies paratuberculosis genetic components in retail cheese curds purchased in Wisconsin and Minnesota by PCR. Molecular and Cellular Probes, 20, 197–202.

27. Rossiter, C.A. and Henning, W.R. (2001) Isolation of Mycobacterium paratuberculosis from thin market cows at slaughter. J Dairy Sci 84(Suppl.), 113.

28. Bell, R.G. and Hathaway, S.C. (1996) The hygienic efficiency of conventional and inverted lamb dressing systems. J Appl Bacteriol 81, 225–234.

29. Meadus, W.J., Gill, C.O., Duff, P., Badoni, M. and Saucier, L. (2008) Prevalence on beef carcasses of Mycobacterium avium ssp. paratuberculosis DNA. Int J Food Microbiol 124, 291–294.

30. Merkal RS, Whipple DL. Inactivation of Mycobacterium bovis in meat products. Appl Environ Microbiol 1980;40: 282–4.

31. Merkal RS, Crawford JA, Whipple DL. Heat inactivation of Mycobacterium avium–Mycobacterium intracellulare complex organisms in meat products. Appl Environ Microbiol 1979;38:831–5

32. Feller, M., K. Huwiler, R. Stephen, et al. 2007. Mycobacterium avium subspecies paratuberculosis and Crohn's disease: a systematic review and meta-analysis. Lancet Infec Dis 7:607-613.

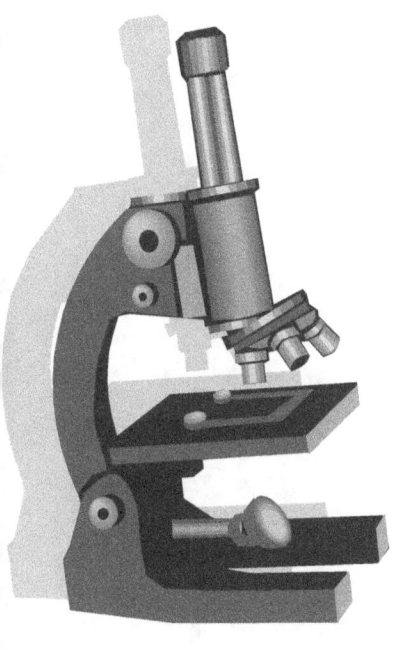

Practicals

PRACTICAL1

COLLECTION, PROCESSING & INOCULATION OF SAMPLES

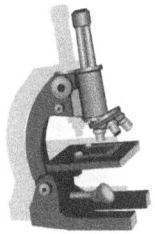

COLLECTION OF SAMPLES (TISSUE/FAECAL SAMPLES)

Intestines and associated mesenteric lymph nodes showing abnormal lesions such as thickening and corrugations of mucosa, and enlargements and oedema of mesenteric and ileocaecal lymph nodes should be collected. To avoid contamination, the faeces should be rinsed from portions of intestinal tract before shipment to the laboratory. Samples were collected in polythene specimen bags kept on ice and were transported to the laboratory. Similarly about 5-10 grams of fresh faecal sample, if possible directly from the rectum or immediately from the ground when voided.

The excess fat and mesenteric attachments in the intestine were trimmed off and intestinal contents were flushed for examination of mucosal lesions and pieces of intestines were preserved in 10% buffered formalin. The mesenteric lymph nodes (MLN) were sectioned longitudinally, examined for gross changes and preserved in formalin. Adjacent intestinal sections and MLN tissues were collected in sterile vials after washing thoroughly with sterile distilled water (DW) for microbiological and molecular studies.

Tissues after fixation were cut into the 2-3 mm thickness, preserved, and embedded in paraffin. Sections of 4-5 μm thickness were prepared from paraffin blocks and stained with Haematoxylin and Eosin (HE).

PROCESSING & INOCULATION

Tissue samples

Digestion/sedimentation method for decontamination of tissues

Approximately 4 g of mucosa from the ileocaecal valve or 4 g of mesenteric node are placed in a sterile blender jar containing 50 ml of trypsin (2.5%). The mixture is adjusted to neutrality using 4% NaOH and pH paper, and stirred for 30 minutes at room temperature on a magnetic mixer.

(94)

The digested mixture is filtered through gauze. The filtrate is centrifuged at approximately 2000–3000 g for 30 minutes. The supernatant fluid is poured off and discarded. The sediment is resuspended in 20 ml of 0.75% HPC and allowed to stand undisturbed for 18 hours at room temperature. The particles that settle to the bottom of the tube are to be used as the inoculum and are removed by pipette without disturbing the supernatant fluid. Alternatively, other methods of decontamination can be used, such as treatment with 5% oxalic acid.

Double incubation method for decontamination of tissues

About 2 g of tissue sample (trimmed of fat) is finely chopped using a sterile scalpel blade or scissors and homogenised in a stomacher for 1 minute in 25 ml 0.75% HPC. Allow the sample to stand so that foam dissipates and larger pieces of tissue settle. Pour tissue homogenate into a centrifuge tube taking care to avoid carry over of fat or large tissue pieces.

Allow settling for 30 minutes then taking 10 ml of the suspension from just above the sediment to a 30 ml tube and incubate for 3 hours at 37°C. Centrifuge for 30 minutes at 900 g, discard supernatant fluid and resuspend pellet in 1 ml antibiotic cocktail containing 100 µg of each of vancomycin, amphotericin and nalidixic acid (VAN). Incubate overnight at 37°C. Use the suspension to inoculate media.

Inoculation of culture media and incubation

Approximately 0.1 ml of inoculum is transferred to each of three slants of Herrold's medium containing mycobactin and to one slant of Herrold's medium without mycobactin. The inoculum is distributed evenly over the surface of the slants. The tubes are allowed to remain in a slanted position at 37°C for approximately 1 week with screw caps loose. The tubes are returned to a vertical position when the free moisture has evaporated from the slants.

(95)

The lids are tightened and the tubes are placed in baskets in an incubator at 37°C. The egg in Herrold's medium contributes sufficient phospholipids to neutralise the bactericidal activity of residual HPC in the inoculum. The other media (Modified Dubos and Middlebrook) do not have this property. Other treatments can be used for sample decontamination, for example oxalic acid at 5%. HPC is relatively ineffective in controlling the growth of contaminating fungi. Amphotericin B (fungizone) was found to control effectively fungal overgrowth of inoculated media (31).

Fungizone may be incorporated in the Herrold's medium at a final concentration of 50 µg per ml of medium. Due to loss of antifungal activity, storage of Herrold's medium containing fungizone should be limited to 1 month at 4°C. The slants are incubated for at least 4 months and observed weekly from the sixth week onwards.

Faecal specimens

No chemical preservative is used. The faecal specimens can be frozen at –70°C.

Suspension and decontamination of faeces

1 g of faeces is transferred to a 50 ml tube containing 20 ml of sterile distilled water. The mixture is shaken for 30 minutes at room temperature. The larger particles are allowed to settle for 30 minutes. The uppermost 5 ml of faeces suspension is transferred to a 50 ml tube containing 20 ml of HPC. The tube is inverted several times to assure uniform distribution and allowed to stand undisturbed for 18 hours at room temperature.

Inoculation of culture media

0.1 ml of the undisturbed sediment is transferred to each of four slants of Herrold's medium, three with mycobactin and one without mycobactin. A smear may be made from the sediment and stained by the Ziehl–Neelsen method. Incubation and observation of slants are same as for tissue specimens.

MEDIA

Herrold's egg yolk medium with mycobactin

For 1 litre of medium: 9 g peptone; 4.5 g sodium chloride; 2.7 g beef extract; 27 ml glycerol; 4.1 g sodium pyruvate; 15.3 g agar; 2 mg mycobactin; 870 ml distilled water; six egg yolks (120 ml); and 5.1 ml of a 2% aqueous solution of malachite green. Measure the first six ingredients and dissolve by heating in distilled water. Adjust the pH of the liquid medium to 6.9–7.0 using 4% NaOH, and test to ensure the pH of the solid phase is 7.2–7.3. Add the mycobactin dissolved in 4 ml ethyl alcohol. Autoclave at 121°C for 25 minutes. Cool to 56°C and aseptically add six sterile egg yolks and sterile malachite green solution. Blend gently and dispense into sterile tubes. It is permissible to add 50 mg chloramphenicol, 100,000 U penicillin and 50 mg amphotericin B.

a.   Use fresh eggs not more than 2 days old from a flock that is not receiving antibiotics. With a brush, scrub the eggs with water containing a detergent. Rinse with water and place the eggs in 70° alcohol for 30 minutes. Dry by inserting between two sterile towels. With sterile rat-tooth forceps, crack one end of the eggshell, making a hole of approximately 10 mm, and remove the egg white with the forceps and gravity. Make the hole larger and break the yolk. Mix the egg yolk by twirling the forceps, and remove the yolk sac. Pour the mixed egg yolk into media.

Modified Dubos's medium

For 1 litre of medium: 2.5 g Difco casamino acids; 0.3 g asparagine; 2.5 g anhydrous disodium hydrogen phosphate; 1 g potassium dihydrogen phosphate; 1.5 g sodium citrate; 0.6 g crystalline magnesium sulphate; 25 ml glycerol; 50 ml of a 1% solution of Tween 80; and 15 g agar. Dissolve each salt in distilled water with minimum heat and make up to 800 ml. Add mycobactin in alcoholic solution at 0.05% (2 mg dissolved in 4 ml ethyl alcohol), heat the medium to 100°C by free-steaming, and then sterilise by autoclaving at 115°C for 15 minutes.

(97)

Cool to 56°C in a water bath, add antibiotics (100,000 U penicillin; 50 mg chloramphenicol; and 50 mg amphotericin B) and serum (200 ml of bovine serum sterilised by filtering through a Seitz 'EX' pad and inactivated by heat at 56°C for 1 hour). The medium is kept thoroughly mixed and then dispensed into sterile tubes. An advantage of this medium is that it is transparent, which facilitates the early detection of colonies.

Modified Middlebrook 7H10

To prepare this medium 19 g Middlebrook 7H10 agar (Difco), 1g Casitone and 5 ml Glycerol are resuspended in 900 ml water, autoclaved at 121°C for 15 minutes and cooled to 58°C. Using an aseptic technique, the following additional ingredients are combined adding the egg yolk last: 50 ml PANTA PLUS (Becton Dickinson), 25 ml Mycobactin J solution (50 µg/ml), 100 ml ADC enrichment (Difco), 250 ml egg yolk. The mixture is thoroughly mixed using a slow swirling action and 10-ml volumes are dispensed into sterile tubes to form slopes. After a sterility check, media are stored at 4°C.

BACTEC 12B vials

The following supplements are added to each vial to give final concentrations of 0.8–1 µg/ml Mycobactin J and a minimum of 16–17% egg yolk in a final volume of 5–6 ml. For the 6-ml volume, 0.1 ml Mycobactin J (50 µg/ml), 0.1 ml PANTA PLUS, 1 ml egg yolk and 0.8 ml water are added. For the 5-ml volume, 0.1 ml Mycobactin J (50 %g/ml), 0.1 ml PANTA PLUS and 0.8 ml egg yolk are added.

Middlebrook 7H9, 7H10 and 7H11 media (Difco)

This media enhanced with mycobactin in the same proportion as for Herrold's medium can also be used. The advantage of this media is that it is transparent, which facilitates the early detection of colonies.

Lowenstein–Jensen medium with or without mycobactin

This media can be used for isolation of organism from fecal samples.

(98)

Reid's synthetic medium

L-Asparagine- 5.0 g, Potassium dihydrogen phosphate (KH2PO4, anhydrous) - 2.0 g, Magnesium sulphate (MgSO4-7H2O)- 1.0 g, Ammonium citrate ([NH4]3 C6H5O7)- 2.0 g, Sodium chloride- 2.0 g, Ferric ammonium citrate- 0.075 g, Dextrose monohydrate B.P.- 10 g, Glycerol B.P. (48 ml)- 60 g, Distilled water to 1000 ml. This media is used for revival of seed cultures of vaccine and johnin producing *Mycobacterium paratuberculosis* strains.

Pratical 2

PREPARATION OF ANTIGENS AND ANTIBODIES

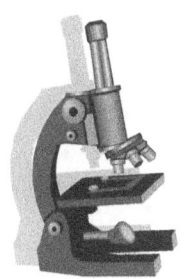

Antigens and antibodies are required for serodiagnostic procedures. The procedures for its preparation are mentioned in detail below.

Preparation of Johnin PPD

Seed culture management

a. Young culture of *Mycobacterium avium subsp. paratuberculosis* strain ATCC 19698 was floated on Watson & Reid's liquid media inside 250 ml flask.

b. Incubated at 37° C. Growth started at 10-15 days with thin film of bacterial growth, appeared throughout the surface of the liquid medium.

c. Seed cultures must not be passaged more than five times

Production culture

a. Watson & Reid's liquid medium, enriched by the addition of trace elements, is used for large scale growth of *M. avium subsp. paratuberculosis*

b. Three liter capacity Hopkins flask containing 1.5 liters of liquid medium are incubated with 3-4 loop full of the thin pellicle growth transferred from seed culture. Incubate at 37° C for 10 to 12 weeks.

c. Flasks showing evidence of contamination and or showing abnormal growth are discarded.

Harvest of production flask

a. After incubation, flasks are steamed for 3 hours at 95^0 to 100^0 C and cells are removed from the cultural filtrate. Before steaming the culture, the pH of the broth is raised to 7.3 using NaOH. Cultural filtrate is collected in sterile flasks with 0.25% phenol as preservative and centrifuged at 3000 rpm for 1 hour at 4° C. Supernatent is passed through sterilized Seitz KS 50 (0.5 micron) and EKS (0.2 micron).

(103)

b. Protein in the culture filtrate is precipitated slowly by adding 4% trichloroacetic acid. The precipitate is slowly allowed to settle overnight at room temperature (pH less than 1.0).

c. Next morning the supernatant is discarded and the precipitate which is having protein is centrifuged at 3000 rpm for 20-30 minutes. Discard the supernatant.

d. The precipitated protein was washed in a solution of 5% NaCl (pH 3). The washing step is repeated three to four times until the pH is 2.7.

e. Stirring the minimum of alkaline solvent dissolves the washed protein. The pH should be about 6.6 to 6.7 and the fluid is clear dark brown. This is the pre - concentrate.

f. Pre- concentrate is centrifuged for 15 minutes at 3000 rpm or higher to remove all the insoluble material and is diluted with an equal amount of glucose buffer solution (R-30). One ml of the sample is used to estimate protein content. The diluted pre – concentrate is kept in the dark at 4° C as a concentrate. The concentrate solution will be diluted to issue strength i.e. 1 mg per ml with glucose buffer solution (R-31) and distilled water.

Potency testing

Each issue is assayed in sensitized guinea pig and cattle to ensure that the potency is equivalent to that of the National or International Standard preparations in appropriate dilution and exhibit required allergic reaction.

Sterility and safety test

The finished product is tested for aerobic and anaerobic contamination, absence of living mycobacteria in appropriate media and freedom from toxicity in guinea pigs and cattle.

Bottling and storage

(104)

1 ml of the product is filled in 2ml capacity ampoule. The ampoule is sealed, labeled and stored at 4°c. The antigen is expected to retain its potency for 2 years if stored at 4°C.

Preparation of sonicated antigen of MAP

The protoplasmic antigen was prepared from *M.a.paratuberculosis*. Culture was grown in the Middlebrook H 10 synthetic media for six weeks at 37^0 C and was killed by autoclaving (120^0 C, 15 lbs pressure) for 5 minutes before harvesting the bacterial cells on Whatman No1 filter paper. The bacterial cells were washed thrice in sterile phosphate buffer solution (PBS; 50 mM phosphate, 150 mM NaCl, pH 7.4) by centrifugation at 5000 rpm for 30 minute. About 5 g of washed cells were resuspended in 10 ml of PBS containing 0.2 mM phenyl methyl sulfonyl fluoride (PMSF) (Himedia laboratories Pvt. Ltd, Mumbai) and sonicated at 16 micron amplitude for 3 minute for 15 cycles with a interval of 2 minute between every 3 cycles (Soniprep 150, Sanyo, Japan). Sonicated preparation was centrifuged (REMI 24: Mumbai, India) at 12000 rpm for 1 hour at 4^0 C. The supernatant was filtered through 0.22 μm membrane filters (MILLEX GV, MILLIPORE, Ireland). The protein content was estimated by Lowry [1] method. The sonicated antigen was aliquoted in small eppendorfs and stored at -20^0C for its use in agar gel immunodiffusion and ELISA (capture antigen).

Preparation of adsorbing antigen of *M. phlei*:

The adsorbing antigen was prepared from *M.phlei*. Culture was grown in the Middlebrook H 10 synthetic media for six weeks at 37^0 C and was killed by autoclaving (120^0 C, 15 lbs pressure) for 5 minutes before harvesting the bacterial cells on Whatman No1 filter paper. The bacterial cells were washed thrice in sterile phosphate buffer solution (PBS; 50 mM phosphate, 150 mM NaCl, pH 7.4) by centrifugation at 5000 rpm for 30 minute.

(105)

About 5 g of washed cells were resuspended in 10 ml of PBS containing 0.2 mM phenyl methyl sulfonyl fluoride (PMSF) (Himedia laboratories Pvt. Ltd, Mumbai) and sonicated at 16 micron amplitude for 3 minute for 15 cycles with a interval of 2 minute between every 3 cycles (Soniprep 150, Sanyo, Japan). Sonicated preparation was centrifuged (REMI 24: Mumbai, India) at 12000 rpm for 1 hour at 4^0 C. The supernatant was filtered through 0.22 µm membrane filters (MILLEX GV, MILLIPORE, Ireland). The protein content estimated was by the Lowry [1] method.

The sonicated antigen referred to as adsorbing was aliquoted in small eppendorfs and stored at -20^0C for its use in ELISA (adsorption antigen).

Raising hyperimmune serum against MAP antigens in rabbits:
Hyperimmune serum was raised as per the methods described previously with suitable modifications [2,3]. The antigen consisting of a mixture of 100 mg of whole cells, 5 mg of sonicated antigen and 2 ml of sonicated sediment was mixed with an equal amount of Freund's incomplete adjuvant (Difco, USA). Each of two rabbits was inoculated with 0.5 ml each of the mixture subcutaneously at weekly intervals for 6 weeks, respectively. Antibody titer was monitored by agar gel immunodiffusion (AGID) test after fifth injection. Rabbits were bled one week after the last injection.

Reference

1. Lowry, O.H., Rosebrough, N.J., Farr, A.L and Randall, R.J. (1951). Protein measurement with folin phenol reagent. *J. Biol. Chem.*, **193**: 265-275

2. Castelnuovo, G., Yamanaka, S., Zeppis, M., and Dotti, E. (1969). The protein components of *M. phlei* fractionation procedures. *Tubercle. Lond.* **50**: 194-202.

(106)

3. Johle, D. K., (2008). Experimental paratuberculosis in rabbits and mice. Ph.D thesis submitted to Indian Veterinary research Institute, Izatnagar.

Practicals 3

DIAGNOSIS OF MAP INFECTION IN ANIMALS

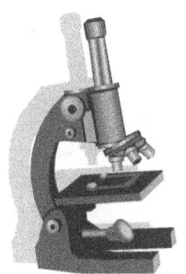

Conventional methods

a. **Ziehl's Neelsen's staining**

Adjacent sections of intestine and mesenteric lymph node were stained with Ziehl-Neelsen's method. Carbol fuschin was poured over the sections, heated with flame until steams comes out and left for 15 min. Then sections were washed and decolorised with 1% acid alcohol until the sections turn pale pink color and then rinsed with water. The decolorised sections were counterstained with methyelene blue for 1 min and washed. The dried sections were dehydrated through alcohol and cleared in xylene, mounted in DPX and were examined under the microscope.

Serological tests

1. Agar gel immunodiffusion test

Preparation of gel and sample loading

One percent agarose gel was prepared in PBS (pH 7.4) containing sodium azide (0.02%). Depending on the number of sample to tested, gel was either poured into the micro slides or petridishes or allowed to solidify at 4^0 C for 1 hour in humid chamber. Wells of 5 mm diameter were punched out in a hexagonal pattern of six peripheral wells for sera and one centre well for the antigen maintaining equi distance of 5 mm between them.

The central well was filled with sonicated antigen with optimum concentration (1 mg/ml). The test sera were charged in duplicate in six peripheral wells and incubated at 4^0C overnight in humid chamber. On each plate, in one set of well positive and negative sera in duplicate were always included as control. Gels were examined after 24 and 48 hr and in suspected cases after 72 hr.

A weak line was scored 1+ and appearance of one or more lines as strong as that of positive serum was scored as 2+.

(111)

Sera testing positive in first instance were reconfirmed for specificity by charging the sera alternatively with know positive serum. Absence of precipitation line was recorded as negative test result.

2. Absorbed ELISA

Optimization of test reagents

For optimization of capture antigen, absorbing antigen and serum dilution in absorbed ELISA, a checkerboard titration should be performed.

Procedure for absorbed ELISA:

a. All the wells in a microtitre plate (NUNC, Maxisorp) were coated with 100 µl of optimized capture antigen (1.2 B preparation of sonicated antigen) in coating buffer and incubated at 37ºC for one hours

b. Unbound antigen in the plate was washed thrice with washing buffer PBS with 0.5% Tween-20 (PBST).

c. Blocking of unbound site was done by adding 100 µl blocking buffer (2% BSA in PBST) and incubated at 37ºC for 1 hour.

d. Plate was washed thrice with wash buffer.

e. The sera samples were incubated with absorbed antigen (1.2 C preparation of sonicated M. phlei antigen) for 30 min following the method of Cox [1] with suitable modification.

f. 100 µl of each sera was added in duplicate wells and incubated for one hour at 37ºC

g. Plate was washed thrice with wash buffer.

h. 100 µl of secondary antibody with HRPO conjugate depending upon the serum sample was added per well and incubated for one hour at 37ºC.

i. Plate was washed thrice with wash buffer.

(112)

j. 100 µl 1x TMB solution (Genei, Bangalore) was added per well and the plate was covered with aluminium foil and observed for colour development

k. After 10-20 minute the reaction was stopped by adding 100 µl of 1 M sulphuric acid per well and reading was taken at 450 nm in an ELISA reader

Determination of S/P values and cut-off value for ELISA positivity

The absorbance values obtained for sera of all experimental animal were expressed and calculated in terms of sample to positive ratio using following formula

$$\text{S/P value (\%)} = \frac{\text{Sample OD} - \text{Known negative control mean}}{\text{Known positive control mean} - \text{Known negative control mean}} \times 100$$

The known negative control mean was calculated by taking the mean of absorbance of all animals which were negative by both PCR and AGID.

The cut-off value for ELISA positivity in the sample animals was calculated following the method described previously with slight modification [2,3].

3. Complement fixation test

The CF test has been the standard test used for cattle for many years. The CF test works well on clinically suspect animals, but does not have sufficient specificity to enable its use in the general population for control purposes. Nevertheless, it is often demanded by countries that import cattle. A variety of CF test procedures are used internationally. An example of a microtitre method for performing the CF test is as follows:

a. The antigen is an aqueous extract of bacteria from which lipid has been removed

(113)

b. All sera are inactivated in the water bath at 60°C for 30 minutes and diluted at 1/4, 1/8 and 1/16. A positive control serum and a negative control serum should be included on each plate. The following controls are also prepared: antigen control, complement control and haemolytic system control.

c. Reconstituted, freeze-dried complement is diluted to contain six times H50 (50% haemolysing dose) as calculated by titration against the antigen.

d. Sheep erythrocytes, 2.5%, are sensitized with 2 units of H100 haemolysin.

e. All dilutions and reagents are prepared in calcium/magnesium veronal buffer; 25 μl volumes of each reagent are used in 96-well round-bottom microtitration plates.

f. Primary incubation is at 4°C overnight and secondary incubation is at 37°C for 30 minutes.

g. *Reading and interpreting the results:* Plates may be left to settle or centrifuged and read as follows: 4+ = 100% fixation, 3+ = 75% fixation, 2+ = 50% fixation, 1+ = 25% fixation and 0 = complete haemolysis.

The titre of test sera is given as the reciprocal of the highest dilution of serum giving 50% fixation. A reaction of 2+ at 1/8 is regarded as positive. Results should be interpreted in relation to clinical signs and other laboratory findings.

4. Indirect immunoperoxidase test

Preparation of silanised slides

Frosted slides were coated with 3-amino propyl triethoxy silane (APES) as per the method described previously [4, 5].

(114)

1.	Slides were cleaned by immersing 2% labolene detergent in pre warmed distilled water (DW) for 30 min after placing in slide rack.

2.	Slides were rinsed thoroughly in DW followed by acetone and then air dried.

3.	Slides were then immersed in 2% APES solution prepared in acetone for 30 min.

4.	Slides were rinsed in acetone, washed in DW, air dried at 37oC and stored in dust free conditions in card board boxes.

Indirect immunoperoxidase staining

Indirect immunoperoxidase staining was carried out as per the method previously described [6].

1.	5-8 micron thin sections were cut and taken onto silanised slides.

2.	Slides were dried at room temperature for 30 min followed by incubation at 56oC for 30 min and stored at room temperature till use.

3.	Deparaffinisation of sections was achieved in two chages of xylene for 15 min each, followed by rehydration through descending grades of alcohol starting from absolute to 70% alcohol and finally to water.

4.	**Heat induced epitope retrieval**: Slides were placed in plastic coplin jar immersed in antigens retrival solution (Vector lab) in jar containing the same solution. Microwave the slides at 900W for 10-15 min. Slides were cooled at room temperature for 30 min, washed in 2 changes of phosphate buffered saline (PBS) for 5 min each (avoid drying of slides while cooling).

5. Buffer on slides were drained off and incubated for 30 min in 3% H2O2 in 80% methanol for blocking the endogenous peroxidase.

6. Following washing in PBS for 5 min, unreactive sites on sections were blocked with 5% normal goat serum (immunohistochemical grade) for 1 hr at 37oC.

7. Slides were not washed rather excess serum was blotted prior to incubation overnight at 4oC with 1:10 diluted polyclonal rabbit hyper immune serum.

8. Sections were washed in 3 changes of PBS for 5 min each, and incubated with 1:200 diluted antirabbit IgG peroxidase conjugate (Sigma, USA).

9. Following washing with 2 changes of PBS for 10 min each, sections were incubated with freshly prepared 0.05% 3, 3' diamino benzidine (DAB) (Sigma, USA) and 0.015% H2O2 in PBS, pH 7.2 for 3-5 min until sections turned light brown in colour. A further reaction was stopped by washing sections in DW and counterstained with Vector

    haematoxylin for 1-5 min.

10. Dip the slides 10 times in acid rinse solution (2 ml glacial acetic acid and 98 ml DW to eliminate faint blue background staining on glass slides and followed by 10 dips in tap water.

11. Incubate slides in bluing solution (1.5 ml NH4OH and 98.5 ml 70% alcohol) for 1 min followed by 10 dips in tap water.

12. Slides were dehydrated in ascending grade of alcohols, cleared in xylene and mounted with Vectamount permanent mounting medium.

    5. Cell mediated immunity

    •

- **Johnin skin test**

The test was carried out by the single intradermal inoculation of 0.1 ml of Johnin purified protein derivatives (PPD, 1mg/ml; BP Division, IVRI, Izatnagar) at the shaven site on one side of the neck.

The skin thickness was measured with calipers before inoculation and at 24 and 48 hr after injection. Animal showing double fold increase in the skin thickness after 48 hr were regarded as reactors as per the manufacturer's instruction

- **Gamma interferon assay**

The assay is based on the release of gamma interferon from sensitized lymphocytes during an 18–36-hour incubation period with specific antigen (avian purified protein derivative [PPD] tuberculin, bovine PPD tuberculin or johnin).

The quantitative detection of bovine gamma interferon is carried out with a sandwich ELISA that uses two monoclonal antibodies to bovine gamma interferon. A commercial diagnostic test based on the detection of gamma interferon has been developed for the diagnosis of bovine tuberculosis. The method and test materials needed are fully described in the instructions accompanying the commercial kit. This test has not been validated by the manufacturer (Prionics, Switzerland) for the diagnosis of paratuberculosis. As such, results derived from this assay are frequently difficult to interpret because there is no agreement with respect to the interpretation criteria and types and amounts of antigens used to stimulate blood lymphocytes. In cattle the reported specificity of the test varied from 94% to 67% depending on the interpretation criteria.

6. **Molecular methods**

Extraction of DNA from fresh tissue/fecal samples

Extraction of DNA from tissue sample was performed as per the

(117)

method described previously with slight modification [7,8].

1. One gram of small intestine and lymph node tissues was homogenized in 1 ml of TE buffer and left undisturbed for 30 min.

2. 500μl supernatant was transferred in another microcentrifuge tube, added with lysozyme (10 mg/ml) and incubated at 37oC for 2 hr.

3. Sample were added with 10% SDS and proteinase K (2mg/ml) and incubated at 56oC overnight.

4. After incubation, the sample mixed with 0.4 volume of 5 M potassium acetate and mixture placed in -20oC for 10 min.

5. The samples were centrifuged at 13,000 rpm for 10 min and supernatant was transferred to another tube.

6. The DNA was purified from the supernatant by adding an equal volume of tris saturated phenol: chloroform: isoamyl-alcohol inverting the tubes (15-20 times) and centrifuging at 13,000 rpm for 10 min.

7. The phenolic extraction was repeated two to three times until the aqueous phase and interphase became clear.

8. The DNA was precipitated from aqueous phase by addition of one tenth volume of 3 M sodium acetate and 2 volumes of absolute alcohol.

9. The tubes were inverted 15-20 times and kept at -20oC for 2 hr to allow DNA precipitation. The DNA was sedimented by centrifugation at 13,000 rpm for 30 min.

10. The supernatant was discarded and DNA pellet was washed with 0.5 ml of 70% alcohol.

11.   The pellet was air dried, resuspended in 25 µl of DNase free water and stored at -20oC for further use.

DNA probes and polymerase chain reaction

DNA probes are being developed that offer a means of detecting *M. paratuberculosis* in diagnostic samples and of rapidly identifying bacterial isolates. They have been used to distinguish between *M. paratuberculosis* and other mycobacteria. McFadden *et al.* have identified a sequence, termed IS*900*, which is an insertion sequence specific for *M. paratuberculosis*.

It has been reported that a small number of isolates other than *M. paratuberculosis* have produced amplified products the same size as expected from *M. paratuberculosis*. A restriction enzyme digest may be applied to positive IS900 products to confirm that their sequence is consistent with *M. paratuberculosis*.

The identifications of new DNA sequences considered to be unique to *M. paratuberculosis*; ISMav2, f57, and ISMap02 sequences, offer additional tools for rapid identification of this organism using the polymerase chain reaction (PCR) technology. The restriction enzyme analysis of IS*1311*, an insertion sequence common to *M. avium* and *M. paratuberculosis* can be used to distinguish between these species and for typing of ovine, bovine and bison strains of *M. paratuberculosis*.

The use of IS*900* as a DNA probe for specific identification of *M. paratuberculosis* in faecal samples from cattle by PCR has been reported (52). Commercial diagnostic PCR tests for the detection of *M. paratuberculosis* in milk and faecal samples have been developed.

Examples of some gene and their primers

PCR primer	Nucleotide sequence	Target gene	Annealing positions	Product length
IS900 F	5'-GGGTTGATCTGGACAATGACGGTTA-3'	IS900	202-226	572
IS900 R	5'-AGCGCGGCACGGCTCTTGTT-3'	IS900	751-771	
f57 F	5'-CCTGTCTAATTCGATCACGGACTAGA-3'	F57	151-176	432
f57 R	5'-TCAGCTATTGGTGTACCGAATGT-3'	F57	558-580	
251 F	5'GCAAGACGTTCATGGGAACT3'	251	-	203
251 R	5'GCGTAACTCAGCGAACAACA 3'	251	-	

PCR amplification protocol

The bacterial genomic DNA isolated from fresh tissue and feacal sample were used as template for amplification of the sequences. The amplification was achieved in 25 µl reaction mixture.

Composition of primary PCR mixture (25µl reaction)

DNA template	- 1.0 µl
Forward primer	- 1.0 µl
Reverse primer	- 1.0 µl
dNTP's (200 µM each)	- 5.0 µl
MgCl2 (1.5 mM)	- 2.5 µl
Taq polymerase	- 0.5 µl
10 X PCR buffer	- 2.5 µl
Distilled water	- 11.5 µl

Total	25 µl

(120)

PCR cycling conditions

Amplification was carried out in a thermal cycler (Eppendorf Master cycler, Hamburg, Germany) with following cycling conditions. The cycling conditions vary according to the primers and gene. The PCR cycling condition for *IS900* gene was given as example here

Stage of cycle	No. of cycle	Temperature	Duration
Initial denaturation	1	94°C	5 min
Denaturation	40 each	94°C	45 sec
Annealing	40 each	68°C	45 sec
Elongation	40 each	72°C	45 sec
Final elongation	1	72°C	10 min
Holding		4°C	

A PCR mixture with water as template was used as negative control and bacterial DNA of MAP reference strain ATCC 19698 was used as positive control.

a. **Analysis of PCR products**

The PCR products were analysed by agarose gel electrophoresis in a submarine horizontal electrophoresis unit (Horizon-58; Life Technologies). 2.0% agarose gel was prepared by boiling the analytical grade agarose (Low EEO, Genei, Bangalore) in 1X TAE buffer in a microwave oven. After cooling to about 50oC, ethidium bromide (Amresco, USA) was added to the agarose solution to obtain a final concentration of 0.5 μg/ml.

(121)

The melted agarose was poured on the gel-casting platform and submerged with sufficient quantity of electrophoresis buffer (1X TAE). The amplified products were mixed with 6X-bromophenol blue loading dye (Promega) and loaded into wells with micropipette.

Electrophoresis was performed at 70volts/cm and progress of mobility was monitored by the migration of dyes. The DNA migration and resolution pattern was examined by UV transillumination and documented.

Reference

1. Cox, J.C., D.P Drane, S.L. Jones, S. Ridge and A.R. Milner (1991). Development and evaluation of a rapid absorbed enzyme immunoassay test for the diagnosis of Johne's disease in cattle. Aust. Vet. J., **68**: 157-160

2. Yokomizo, Y., Merkal, R.S. and Lyle, P.A.S. (1983). Enzyme-linked immunosorbent assay for detection of bovine immunoglobulin G1 antibody to protoplasmic antigen of Mycobacterium paratuberculosis. Am. J. Vet. Res., **44**: 2205-2207.

3. Rajukumar, K., Tripathi, B.N., Kurade, N.P. and Parihar, N.S., (2001). An enzyme linked immunosorbent assay using immuno-affinity-purification antigen in the diagnosis of caprine paratuberculosis and its comparison with conventional ELISAs. *Vet. Res. Commun.,* **25:** 539-553.

4. Herrington, C. S., and Mc-Gee, J. O. D. (1992). Diagnostic molecular Pathology: A practical approach, Vol. 1, pp80, Oxford University Press, Oxford.

5. Tripathi (2007). Immunopathological and molecular techniques in Diagnostic Pathology. 1st edn Officer Incharge, Communication centre, Indian Veterinary research Institute, Publication. Ltd, India.

6. Kumar, A.A., Tripathi, B.N. and Julhe, D.K. (2006). Immunohistochemical demonstration of mycobacterial antigen in sheep experimentally infected with mycobacterium avium subsp. paratuberculosis. *Indian J. Vet. Pathol.*, **30(2):** 1-4.

7. Sivakumar, P., Tripathi, B.N. and Singh, N. (2005). Detection of mycobacterium avium subsp. paratuberculosis in intestinal and lymphnode tissues of water buffaloes (*Bubalis bubalis*) by PCR and bacterial culture. *Vet. Microbiol.*, **108:** 263-270.

8. Tripathi, B.N., Sivakumar Periasamy, Paliwal, O.P. and Singh, N (2006). Comparison of IS*900* tissue PCR, bacterial culture, johnin and serological test for diagnosis of naturally occurring paratuberculosis in goats. *Vet. Microbiol.*, **116:** 129-137.

ANNEXURE

Solutions for agarose gel electrophoresis

Tris-acetate - EDTA (TAE) buffer (50x)

Tris base	24.2 g
Glacial acetic acid	5.71 ml
0.5 M EDTA	10 ml

Volume made upto 100 ml with double distilled water.

Sterilized by autoclaving for 20 min at 15 psi on liquid cycle and stored at room temperature.

Ethidium bromide solution (10 mg/ml)

10 mg of ethidium bromide dissolved in 1 ml of autoclaved distilled water and vortexed till it get dissolved and stored protected from light.

6x Gel loading dye

Bromophenol blue	0.25%
Xylene cyanol	0.05%
Glycerol	60%

Phosphate buffered saline (pH 7.4)

NaCl	8 g
KCl	0.2 g
KH_2PO_4	0.2 g
$Na_2 HPO_4$ (anhydrous)	1.14 g (or)
$Na_2 HPO_4.2H_2O$	

Dissolved and volume made to 1000 ml with distilled water.

Tris-EDTA (TE) buffer (pH 8.0)

Tris-HCl (pH 8.0)	10 mM
EDTA (pH 8.0)	1 mM

Proteinase K

20 mg/ml in TE buffer

TBE buffer (10xstock solution)

Tris base	108 g
Boric acid	55 g
0.5 M EDTA	40 ml

Dissolved in 800 ml of autoclaved distilled water by stirring adjust the volume to 1 litre with autoclaved distilled.

1M Tris-HCl (pH 7.4 and 8.0)

Tris base	121.1 g
Distilled water	upto 1000 ml

Adjust the pH to the desired volume by adding conventional HCl.

Dispense into aliquots and sterilize by autoclaving at 15 psi for 20 min.

Reagents for ELISA

Coating buffer (0.5 M carbonate/biocarbonate buffer, pH 9.6) 1 litre.

Sodium carbonate	1.59 g
Sodium bicarbonate	2.93 g

Stored at 4°C for maximum of 15 days.

Phosphate buffered saline (PBS) pH 7.4 (50 mM phosphate, 150 mM NaCl, pH 7.4)

Disodium hydrogen phosphate 2H$_2$O	1.16 g

(128)

Sodium chloride	8 g
Potassium dihydrogen phosphate	0.2 g
Potassium chloride	0.2 g
Distilled water	1000 ml

Washing buffer (PBST, pH 7.4)

PBS (pH 7.4)	1000 ml
Tween-20	0.5 ml

Dilution buffer

PBST	100 ml
BSA	1.0 g

Prepared just before use

Conjugate

Anti sheep IgG perioxidase (Sigma USA)	1 part
PBS	5000 parts
Anti goat IgG perioxidase (Sigma USA)	1 part
PBS	5000 parts

Enzyme activity blocking solution (5N H$_2$SO$_4$)

Sulphuric acid	14.2 ml
Distilled water	100 ml

www.ingramcontent.com/pod-product-compliance
Lightning Source LLC
Chambersburg PA
CBHW081259170526
45165CB00011B/3357